Ernst Schering Foundation Symposium
Proceedings 2007-4
Oncogenes Meet Metabolism

Ernst Schering Foundation Symposium
Proceedings 2007-4

Oncogenes Meet Metabolism

From Deregulated Genes
to a Broader Understanding
of Tumour Physiology

G. Kroemer, D. Mumberg, H. Keun, B. Riefke,
T. Steger-Hartmann, K. Petersen
Editors

With 60 Figures

Series Editors: G. Stock and C. Klein

Library of Congress Control Number: 2008925641

ISSN 0947-6075

ISBN 978-3-540-79477-6 Springer Berlin Heidelberg New York

This work is subject to copyright. All rights are reserved, whether the whole or part of the material is concerned, specifically the rights of translation, reprinting, reuse of illustrations, recitation, broadcasting, reproduction on microfilms or in any other way, and storage in data banks. Duplication of this publication or parts thereof is permitted only under the provisions of the German Copyright Law of September 9, 1965, in its current version, and permission for use must always be obtained from Springer-Verlag. Violations are liable for prosecution under the German Copyright Law.

Springer is a part of Springer Science+Business Media
springer.com

© Springer-Verlag Berlin Heidelberg 2008

The use of general descriptive names, registered names, trademarks, etc. in this publication does not emply, even in the absence of a specific statemant, that such names are exempt from the relevant protective laws and regulations and therefor free for general use. Product liability: The publisher cannot guarantee the accuracy any information about dosage and application contained in this book. In every induvidual case the user must check such information by consulting the relevant literature.

Cover design: design & production, Heidelberg
Typesetting and production: le-tex publishing services oHG, Leipzig
21/3180/YL – 5 4 3 2 1 0   Printed on acid-free paper

Preface

There is strong evidence for a common metabolic phenotype associated with cancer, observed both in vivo and in vitro, across species, and across a wide range of primary and secondary tumour sites. Already in 1920s, Otto Warburg described the phenomenon of "aerobic glycolysis", the apparently greater tendency of tumour cells to convert glucose to lactate in the presence of normal oxygen conditions. At his time, Warburg's hypothesis that cancer was caused by altered metabolism found no wide acceptance even though other observations that growth of tumour cells in culture is often unusually dependent on the availability of common substrates, such as glutamine, arginine, methionine and cysteine, support the idea of a tumour metabolic phenotype. To what extent this metabolic phenotype of cancer is causal or consequential to carcinogenesis and disease progression is still not clear, but important evidence exists to suggest that it confers selective growth advantages to transformed cells.

While the debate still continues as to the significance of the Warburg effect, at least one aspect of the phenotype, namely, increased glucose uptake, is already being exploited clinically by PET imaging. Although the Warburg effect has been demonstrated and confirmed in most human tumours, the advent of molecular biology and the discovery of oncogenes and tumour suppressor genes in the 1970s have shifted the scientific interest in tumour metabolism towards the search for the genetic basis for cancer. To date, the focus in the cancer molecular profiling community has been in serum proteomics for diagnostic markers, and in tissue/cell transcriptomics for prognostic markers. However, metabolites have a key advantage as biomarkers: they are highly translatable from laboratory work to the clinic. This is in part due to the fact that a metabolite is the same chemical entity irrespective of whether it is observed in a cell, organelle, tissue or biofluid, whichever individual, sex or species is being observed at the time. The same is not true for genes or proteins, which undergo alternative splicing and translational modifications in addition to sequence variation. Today technologies are available to rapidly analyse broad varieties of metabolites in various tissues and body fluids (metabonomics) and interpret the data. Additionally, it could be shown that a variety of oncogenes exert their transforming activities largely by modulating central metabolic pathways, such as glycolysis. Particularly, those oncogenes protecting cancer cells from naturally occurring programmed cell death (apoptosis) appear to act predominantly by ensuring sufficient nutrient supply and energy production for the malignant cancer cells. Thus, we are starting to unravel how oncogenic signal transduction is connected to the metabolism, survival and growth of tumour cells. Obviously, these findings will have tremendous consequences for the understanding of the molecular mechanisms leading to cancer. Furthermore, they open new opportunities for the development of new therapeutic drugs and diagnostic tools for the treatment of cancer, as exemplified by the recent finding that LDH (lactate dehydrogenase) may be a predictive marker for the response of tumours to anti-angiogenic therapies.

A long-awaited promise of the post-genomic era was the use of biomolecular profiling, particularly genetic profiling, to tailor the therapy of each individual to their specific needs and susceptibilities. This goal of personalized medicine has already begun in oncology by the selective application of drugs such as Herceptin that only benefit a sub-population of patients based on the genetic makeup of their tumour (over-expression of HER2). Metabonomics has enormous potential in this area, not only because metabolic biomarkers can act as phenotypic indicators for expression of genetic differences as described above, but also because essentially the same analytical protocol and platform used to discover a metabolic biomarker in an experimental model can be applied in subsequent preclinical efficacy and safety studies, as well as

Preface

in later clinical trials. Therefore metabonomics, whether based on mass spectrometry or nuclear magnetic resonance, can be considered as an ideal technology to detect translational biomarkers.

The Workshop "Oncogenes Meet Metabolism—From Deregulated Genes to a Broader Understanding of Tumour Physiology" was organized in order to discuss the recent advances and controversies in this fast-moving research area. We tried to bring together many of the internationally recognized experts who, through a variety of approaches, have made seminal contributions, thus leading to major strides forward. We are grateful to all of them for their excellent presentations and lively discussions, and also for their contributions to this book. We are convinced that the proceedings of the workshop will allow a better understanding of important aspects of the metabolism of tumours and will help in the future development of more effective and selective cancer diagnostics and treatments.

Finally, we would like to express our gratitude to the Ernst Schering Foundation for its generous support and superb organization, which allowed us to hold this workshop in the best possible conditions.

Björn Riefke
Dominik Mumberg
Guido Kroemer
Hector Keun
Kirstin Petersen
Thomas Steger-Hartmann

Contents

Mitochondria and Cancer
P. Rustin, G. Kroemer 1

Role of the Metabolic Stress Responses of Apoptosis
and Autophagy in Tumor Suppression
E. White 23

The Interplay Between MYC and HIF in the Warburg Effect
C.V. Dang 35

Using Metabolomics to Monitor Anticancer Drugs
Y.-L. Chung, J.R. Griffiths 55

Biomarker Discovery for Drug Development
and Translational Medicine Using Metabonomics
H.C. Keun 79

Pyruvate Kinase Type M2: A Key Regulator
Within the Tumour Metabolome
and a Tool for Metabolic Profiling of Tumours
S. Mazurek 99

Molecular Imaging of Tumor Metabolism and Apoptosis
U. Haberkorn, A. Altmann, W. Mier, M. Eisenhut 125

Minimally Invasive Biomarkers for Therapy Monitoring
P. McSheehy, P. Allegrini, S. Ametaby, M. Becquet, T. Ebenhan,
M. Honer, S. Ferretti, H. Lane, P. Schubiger, C. Schnell,
M. Stumm, J. Wood 153

Use of Metabolic Pathway Flux Information
in Anticancer Drug Design
L.G. Boros, T.F. Boros 189

Cancer Diagnostics Using H-NMR-Based Metabonomics
K. Odunsi 205

Human Metabolic Phenotyping and Metabolome
Wide Association Studies
E. Holmes, J.K. Nicholson 227

Defining Personal Nutrition and Metabolic Health
Through Metabonomics
S. Rezzi, F-P.J. Martin, S. Kochhar 251

List of Editors and Contributors

Editors

Kroemer, G.
INSERM, U848, Institut Gustave Roussy, Pavillion de Recherche 1,
39, rue Camille Desmoulins, 94805 Villejuif, France
(e-mail: kroemer@igr.fr)

Mumberg, D.
Tumour Biology and Translational Oncology Research,
Bayer Schering Pharma AG, 13342 Berlin, Germany
(e-mail: dominik.mumberg@bayerhealthcare.com)

Keun, H.
Imperial College Faculty of Medicine, Biomolecular Medicine,
Division of Surgery, Oncology, Reproductive, Biology and Anaestetics,
Sir Alexander Fleming Building, Exhibition Road, South Kensington,
London SW7 2AZ, UK
(e-mail: h.keun@imperial.ac.uk)

Riefke, B.
GDD – Nonclinical Drug Safety, Bayer Schering Pharma AG,
Müllerstaße 178, 13353 Berlin, Germany
(e-mail: Bjoern.riefke@bayerhealthcare.com)

Petersen, K.
Tumour Biology and Translational Oncology Research,
Bayer Schering Pharma AG, 13342 Berlin, Germany
(e-mail: kirstin.petersen@bayerhealthcare.com)

Steger-Hartmann, T.
Nonclinical Drug Safety, Bayer Schering Pharma AG, 13342 Berlin,
Germany
(e-mail: thomas.steger-hartmann@bayerhealthcare.com)

Contributors

Allegrini, P.
Discovery Technologies, Novartis Pharma AG, 4002 Basel, Switzerland

Altmann, A.
Department of Nuclear Medicine, University of Heidelberg,
INF 400, 69120 Heidelberg, Germany
and Clinical Cooperation Unit Nuclear Medicine,
DKFZ and University of Heidelberg, INF 280, 69120 Heidelberg

Ametaby, S.
Animal Imaging Center, Radiopharmaceutical Sciences of ETH,
PSI and USZ, 8093 Zurich, Switzerland

Becquet, M.
Oncology Research, Novartis Pharma AA, 4002 Basel Switzerland

Boros, T.F.
Los Angeles Biomedical Research Institute at Harbour –
UCLA Medical Center RB1,
1124 West Carson Street, Torrance, CA 90502, USA

List of Editors and Contributors

Boros, L.G.
Los Angeles Biomedical Research Institute at Harbour –
UCLA Medical Center RB1,
1124 West Carson Street, Torrance, CA 90502, USA
(e-mail: lboros@sidmap.com)

Chung, Y.-L
St. George's University of London, London UK

Dang, C.V.
Department of Medicine, Cell Biology, Molecular Biology &
Genetics, Oncology, and Pathology, Johns Hopkins University
School of Medicine, Ross Research Building, Room 1032,
720 Rutland Avenue, Baltimore, MD 21205, USA
(e-mail: cvdang@jhmi.edu)

Ebenhan, T.
Animal Imaging Center, Radiopharmaceutical Sciences of ETH,
PSI and USZ, 8093 Zurich, Switzerland

Eisenhut, M
Department of Radiopharmaceutical Chemistry, DKFZ, INF 280,
69120 Heidelberg, Germany

Ferretti, S.
Oncology Research, Novartis Pharma AG, 4002 Basel, Switzerland

Griffiths, J.R.
Cancer Research UK, Cabridge Research Institute, Robinson Way,
Cabridge CB2 0RE, UK
(e-mail: John.Griffiths@cancer.org.uk)

Haberkorn, U.
Department of Nuclear Medicine, University of Heidelberg,
INF 400, 69120 Heidelberg, Germany
and Clinical Cooperation Unit Nuclear Medicine,
DKFZ and University of Heidelberg, INF 280, 69120 Heidelberg
(e-mail: uwe.haberkorn@med.uni-heidelberg.de)

Honer, M.
Animal Imaging Center, Radiopharmaceutical Sciences of ETH,
PSI and USZ, 8093 Zurich, Switzerland

Holmes, E.
Divsion of Surgery, Oncology, Reproductive Biology
and Anaesthetics (SORA), Faculty of Medicine,
Sir Alexander Fleming Building,
South Kensington, London SW7 2AZ, UK
(e-mail: Elaine.holmes@imperial.ac.uk)

Kochhar, S.
Nestlé Research Center BioAnalytical Science, Metabonomics
and Biomarkers, PO Box 44, 1000 Lausanne 26, Switzerland
(e-mail: sunil.kochhar@rdls.nestle.com)

Lane, H.
Oncology Research, Novartis Pharma AG, 2002 Basel, Switzerland

Martin, F.-P.J.
Nestlé Research Center BioAnalytical Science, Metabonomics
and Biomarkers, PO Box 44, 1000 Lausanne 26, Switzerland

Mazurek, S.
Institute for Biochemistry and Endocrinology, Veterinary Faculty,
University of Giessen, Frankfurter Strasse 100, 35392 Giessen, Germany
(e-mail. Sybille.Mazurek@vetmed.uni-giessen.de)

McSheehy, P.
Oncology Research, Novartis Pharma AG, 4002 Basel, Switzerland
(e-mail: paul_mj.mcsheehy@novartis.com)

Mier, W.
Department of Nuclear Medicine, University of Heidelberg,
INF 400, 69120 Heidelberg, Germany
and Clinical Cooperation Unit Nuclear Medicine,
DKFZ and University of Heidelberg, INF 280, 69120 Heidelberg

List of Editors and Contributors

Odunsi, K.
Departments of Gynecologic and Oncology Immunology,
Roswell Park Cancer Institue, Elma and Carlton Streets,
Buffalo, NY 14261, USA
(e-mail: Kunle.Odunsi@Roswellpark.org)

Rezzi, S.
Nestlé Research Center BioAnalytical Science, Metabonomics
and Biomarkers, PO Box 44, 1000 Lausanne 26, Switzerland

Rustin, P
INSERM, U676, Hopital Robert Debre, 75019 Paris, France

Schnell, C.
Oncology Research, Novartis Pharma AG, 2002 Basel, Switzerland

Schubiger, P.
Animal Imaging Center, Radiopharmaceutical Sciences of ETH,
PSI and USZ, 8093 Zurich, Switzerland

Stumm, M.
Oncology Research, Novartis Pharma AG, 2002 Basel, Switzerland

White, E.
Rutgers University, Molecular Biology and Biochemistry,
CABM-Room 140, 679 Hoes Lane, Piscataway, NJ 08854, USA
(e-mail: ewhite@cabm.rutgers.edu)

Wood, J.
S*Bio Pte Ltd, 1 Science Park Road, #05–09, The Capricorn,
Singapore Science Park II, Singapore 117528

Ernst Schering Foundation Symposium Proceedings, Vol. 4, pp. 1–21
DOI 10.1007/2789_2008_086
© Springer-Verlag Berlin Heidelberg
Published Online: 06 June 2008

Mitochondria and Cancer

P. Rustin, G. Kroemer[(✉)]

Institut Gustave Roussy, INSERM, U848, Pavillion de Recherche 1,
39 rue Camille Desmoulins, 94805 Villejuif, France
email: *kroemer@igr.fr*

1	Introduction	2
2	Mitochondrial Control of Apoptosis	3
3	Therapeutic Interventions for the Restoration of Mitochondrial Apoptosis in Cancer Cells	7
4	Reduced Oxidative Phosphorylation and Carcinogenesis	9
5	Hypothetical Links Between Apoptosis Resistance and Anaerobic Glycolysis at the Mitochondrial Membrane	14
References		16

Abstract. Mitochondria contained in cancer cells exhibit two major alterations. First, they are often relatively resistant to the induction of mitochondrial membrane permeabilization (MMP), which is the rate-limiting step of the intrinsic pathway of apoptosis. The mechanisms of MMP resistance have come under close scrutiny because apoptosis resistance constitutes one of the essential hallmarks of cancer. Second, cancer cell mitochondria often exhibit a reduced oxidative phosphorylation, meaning that ATP is generated through the conversion of glucose to pyruvate and excess pyruvate is then eliminated as the waste product lactate. This glycolytic mode of energy production is even observed in conditions of high oxygen tension and is hence called anaerobic glycolysis. Here, we discuss the molecular mechanisms accounting for inhibition of the mitochondrial apoptosis pathway in neoplasia and discuss possible mechanistic links between MMP resistance and anaerobic glycolysis.

1 Introduction

When cells are kept in a glucose-rich milieu and are cultured first in a hypoxic environment and then in a normoxic one, they manifest the so-called Pasteur phenomenon, that is a reduction in glucose consumption, concomitant with a decrease in lactate production and an increase in oxygen consumption. In contrast, cancer cells behave differently and continue high glycolysis and lactate production, even in conditions of high oxygen tension. This phenomenon is referred to as anaerobic glycolysis and was discovered in the 1920s by the late Nobel Prize winner Otto Warburg as the first biochemical hallmark of cancer (Warburg et al. 1924, 1926). Nonetheless, Warburg was unable to demonstrate that the Warburg phenomenon would account for oncogenesis or participate in tumor progression as a causative factor. Indeed, this hypothesis was dismissed, and the study of intermediate metabolism and oxidative phosphorylation (which is decreased in cancer cells) was abandoned with the advent of molecular biology and the discovery of oncogenes and tumor suppressor genes that have captured most if not all of the attention of cancer biologists over the last three decades.

As a result, cancer biologists and medical oncologists have been considering the university courses in which they were taught that mitochondrial metabolism, including the tricarboxylic acid cycle (TCA) and oxidative phosphorylation (OXPHOS), were a useless and time-consuming effort that they discretely abhorred. Nonetheless, there was a sudden and unexpected renaissance of mitochondrial biology when it was discovered that these organelles control cell death (Kroemer et al. 2007; Liu et al. 1996; Zamzami et al. 1996). Indeed, it appears that mitochondrial membrane permeabilization (MMP) is often the decisive event that marks the frontier between survival and death, irrespective of the morphological features of end-stage cell death (which may be apoptotic, necrotic, autophagic or mitotic). In a way, mitochondrial membranes make up the battleground on which opposing vital and lethal signals combat to seal the cell's fate. Local players that modulate the propensity to MMP include the pro- and anti-apoptotic members of the Bcl-2 family (Adams and Cory 2007b), proteins from the mitochondrial permeability transition pore complex (PTPC), as well as a cornucopia of interacting partners including mitochondrial lipids (Zamzami and Kroe-

Mitochondria and Cancer 3

mer 2001). Intermediate metabolites, redox reactions, sphingolipids, ion gradients, transcription factors, as well as kinases and phosphatases, link survival or death signals emanating from distinct subcellular compartments to mitochondria. Thus, mitochondria have the capacity to integrate multiple pro- and anti-apoptotic signals. Once MMP has been triggered, it causes the release of catabolic hydrolases and activators of such enzymes (including those of caspases) from mitochondria. These catabolic enzymes as well as the cessation of the bioenergetic and redox-detoxifying functions of mitochondria finally cause cellular demise, implying that mitochondria coordinate the late stage of cell death. In tumor cells, MMP is inhibited at the level of mitochondria or upstream thereof, at the level of premitochondrial pro-apoptotic signal transduction pathways. Induction of MMP in transformed cells constitutes the goal of anti-cancer chemotherapy (Kroemer et al. 2007).

The purpose of the present review is to briefly discuss the mechanisms of MMP inhibition in tumor cells and to establish hypothetical links between MMP resistance and anaerobic glycolysis.

2 Mitochondrial Control of Apoptosis

Apoptosis is morphologically defined as a type of cell death in which the cell and, in particular, the nucleus shrinks (Kroemer et al. 2005). Chromatin condensation (pyknosis) and nuclear fragmentation (karyorrhexis) are the two hallmarks that define apoptosis. Although there have been attempts to define apoptosis biochemically (for instance, as cell death with caspase activation or cell death with phosphatidylserine exposure on the outer leaflet of the plasma membrane), these attempts have failed, for the simple reason that the alleged specific hallmarks of apoptotic cell death are not truly specific (thus, phosphatidylserine exposure and caspase activation can occur during T cell activation without cell death) (Galluzzi et al. 2007; Kroemer et al. 2005). Given that the morphology of nuclei changes in a much more characteristic (and specific) fashion than that of any other organelle, in apoptosis, at the beginning it was thought that these changes would reflect the essence of the apoptotic process and that the point-of-no-return, the frontier between death and life, would be determined by alterations in nuclear morphology related to degradation of nuclear DNA (chromatinolysis).

This concept was invalidated, initially by a cell-free system in which cellular constituents (organelles and cytosol) were admixed in vitro to recapitulate the process culminating in nuclear alterations (pyknosis, karyorrhexis, chromatinolysis). Using this system, we discovered that the most reproducible way to induce apoptosis in vitro was the following: in a first step, cells were treated with an apoptosis inducer. After a short incubation, the cytosol that contained accumulating MMP inducers was purified. In the second step, this cytosol was then mixed with mitochondria from healthy cells, resulting in MMP and hence the release of pro-apoptotic effector molecules through the permeabilized outer mitochondrial membrane. In the third and final step, these effector molecules were added to healthy nuclei to induce apoptotic changes (Susin et al. 1996, 1997; Zamzami et al. 1996).

Using this system, we purified and identified apoptosis-inducing factor, a caspase-independent death effector that acts on purified nuclei to cause peripheral chromatin condensation and large-scale DNA fragmentation to approximately 50 kbp (Susin et al. 1999). Using a slightly different cell-free system, Xiadong Wang and colleagues purified and identified all the constituents of the postmitochondrial caspase activation pathway, namely, cytochrome c (which leaks out from the mitochondrial intermembrane space), Apaf-1 (an ATP-dependent adaptor) and caspase-9 (the apical caspase of a cascade culminating in the activation of the effector caspases-2, -6 and -7) (Li et al. 1997; Liu et al. 1996; Zou et al. 1997). Importantly, the anti-apoptotic action of the oncoprotein Bcl-2 was mapped by several groups (Kluck et al. 1997; Susin et al. 1996; Yang et al. 1997) at the mitochondrial level, meaning that Bcl-2 interrupts the apoptotic process by sealing the mitochondrial membranes and by preventing MMP.

These results as well as other experiments transposed the nucleocentric world view of apoptosis to a mitochondriocentric one (Fig. 1). The mitochondrion, and in particular, MMP would determine the decision of committing apoptotic suicide, acting as the central integration point of the apoptotic process and then as the coordinator of the catabolic process that leads to ordered cellular dismantling (Kroemer et al. 1995). Obviously, this concept had far-reaching implications for the concep-

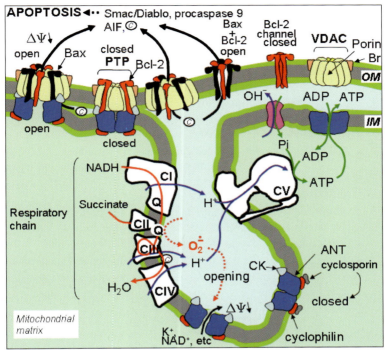

Fig. 1. Mitochondrial pathways to apoptosis. The release of intermembrane space components (such as AIF, or cytochrome *c*) and/or the loss of membrane potential (ΔΨ) as affected by numerous stimuli trigger caspase activation and cell commitment to die through an apoptotic process. Loss of membrane potential and of various matrix cofactors can result from the opening of the mitochondrial permeability transition pore (*PTP*). The balance between the pro- and anti-apoptotic members of the Bcl-2 family controls the opening of the pore. Alternatively, the members of this family can form channels that may also allow for the release of proapoptotic components present in the intermembrane space. PTP is a complex formed between the voltage-dependent anion channel (*VDAC*) of the outer membrane and the adenylate nucleotide translocase (*ANT*) of the inner membrane associated with several additional proteins. *AIF*, apoptosis-inducing factor; *Br*, benzodiazepine-receptor; *CI–CV*, the various complexes of the respiratory chain; *c*, cytochrome *c*; *IM*, inner membrane; *OM*, outer membrane

tion, detection and manipulation of cell death at several levels (Jiang and Wang 2004; Kroemer et al. 2007). This can be easily illustrated by the pathophysiological implications of MMP-mediated cell death control:

- Many different signals can induce (or inhibit) MMP, linking different types of cellular stress and damage to mitochondria. This underscores the potential of mitochondria to function as general cell death sensors and to integrate many distinct lethal triggers (Brenner and Kroemer 2000).

- MMP is not simply induced (or inhibited) by a single class of molecules. Rather, several alternative, complementary and intertwined modes of MMP exist that are mediated by distinct classes of proteins and modulators (Zamzami and Kroemer 2001). This introduces some sort of redundancy into the system that regulates cell death, hence preventing a mutation completely suppressing cell death, an event that would be intrinsically oncogenic.

- When MMP has trespassed a critical threshold, its biochemical consequences possibly encompass further permeabilization of adjacent and distant mitochondria, thereby resulting in a rapid self-amplifying phenomenon, which occurs prominently in an all-or-nothing fashion. This implies that the detection of MMP indeed predicts imminent cell death.

- Once MMP has occurred, it triggers cell death rapidly and efficiently, through a plethora of independent and redundant mechanisms. These include the activation of caspases and caspase-independent death effectors, as well as irreversible metabolic changes at the bioenergetic and redox levels.

- If cytoprotection is the therapeutic goal, it is indispensable to prevent MMP or the upstream events leading to MMP. In contrast, cellular demise cannot be avoided by inhibiting the post-mitochondrial phase of apoptosis, which comprises biochemical changes occurring after the point of no return has been trespassed (postmortem events). This is essential for the design of neuro-, hepato-, nephro- or cardioprotective therapies.

- Pathological MMP contributes to the unwarranted loss of post-mitotic cells in the brain and heart. Pharmacological agents that target specific mitochondrial ion channels or proteins that contribute to MMP may be useful for the therapeutic suppression of acute cell death.

- Cancer cells are often relatively resistant to MMP induction, and the therapeutic induction of MMP constitutes a therapeutic goal in anti-cancer chemotherapy or radiotherapy. The inhibition of MMP-inhibitory proteins (such as Bcl-2-like proteins) can sensitize tumor cells to apoptosis induction (Kroemer et al. 2007). This latter point will be discussed in more detail in the following section.

3 Therapeutic Interventions for the Restoration of Mitochondrial Apoptosis in Cancer Cells

The inhibition of cell death is one of the hallmarks of cancer. Apoptosis is inhibited in carcinomas, sarcomas, melanomas and hematopoietic malignancies, either upstream or at the mitochondria. One of the most prominent examples of apoptosis inhibition acting at the mitochondrial level is the overexpression of anti-apoptotic proteins of the Bcl-2 family such as Bcl-2 or its close homologues $Bcl-X_L$ and Mcl-1 (Adams and Cory 2007a,b). Bcl-2, $Bcl-X_L$ and Mcl-1 are multidomain proteins and carry four distinct Bcl-2 homology (BH) domains that are labeled BH1–BH4. The branch of pro-apoptotic multidomain proteins of the Bcl-2 family comprises Bax and Bak, which both possess BH1, BH2 and BH3 domains, yet lack a BH4 domain. Finally, a vast group of at least a dozen different proteins makes up the so-called BH3-only branch of the Bcl-2 family. Together, these proteins can regulate MMP induction in many instances, and cancer cells can be resistant to MMP stimulation due to the overexpression of anti-apoptotic Bcl-2 proteins or the absence of pro-apoptotic Bcl-2 family proteins (Deng et al. 2007). As a result, one of the most specific therapeutic interventions that can be created, on theoretical grounds, is a specific ligand that inhibits anti-apoptotic or activates pro-apoptotic proteins of the Bcl-2 family. For example, ABT737 has been designed as a specific ligand that inactivates Bcl-2 (and Bcl-XL) (Oltersdorf et al. 2005), and ABT737 derivatives with improved pharmacokinetic properties are currently under clinical evaluation.

Another prominent collection of proteins that mediate MMP (or at least impinge on the probability of MMP induction) are the proteins contained in the so-called permeability transition pore complex (PTPC) (Zamzami and Kroemer 2001). Although the exact composition of the PTPC is still a matter of debate, it appears that this complex involves interactions between hexokinase (HK, an enzyme that catalyzes the initial step of glycolysis in the cytosol), the voltage-dependent anion channel (VDAC, a largely nonspecific pore in the outer mitochondrial membrane), the mitochondrial benzodiazepine receptor (in the outer membrane), the adenine-nucleotide translocator (ANT, the electrogenic antiporter of ATP and ADP on the inner mitochondrial membrane) and cyclophilin D (a prolyl *cis-trans* isomerase located in the mitochondrial matrix). Accordingly, inhibitors of the HK-VDAC interaction (Pedersen 2007), pharmacological components acting on VDAC (Yagoda et al. 2007), ligands of the mitochondrial benzodiazepine receptor (Decaudin et al. 2002), or knockdown of the ANT2 isoenzyme (Le Bras et al. 2006) may have apoptosis-inducing, antineoplastic effects.

A vast collection of pharmacological agents may exert direct MMP-inducing effects on isolated mitochondria, and the exact mode of action of these agents is often incompletely characterized (Costantini et al. 2000; Galluzzi et al. 2006). Prominent MMP inducers include lipophilic cations that enrich in cancer mitochondria (which are often hyperpolarized) and that trigger MMP, presumably through yet-to-be defined interactions with mitochondrial inner membrane proteins and/or lipids. Several among these agents are in preclinical development. On theoretical grounds, such direct MMP inducers may circumvent the apoptosis resistance that characterized transformed cells. We have published several reviews (Costantini et al. 2000; Galluzzi et al. 2006) on this important topic, enumerating the distinct compounds that can trigger MMP in a direct fashion not requiring the cell to generate MMP-inducing signal transducers. This strategy of cell death induction has the obvious advantage of readily bypassing mechanisms of apoptosis resistance that reside in the generation of MMP inducers (such as defects in the p53-dependent pro-apoptotic signal transduction pathway) or that affect the composition of mitochondrial membranes themselves.

4 Reduced Oxidative Phosphorylation and Carcinogenesis

Otto Warburg made the seminal observation that, under aerobic condition, cancer cells paradoxically shift their metabolism from an oxidative metabolism (through the mitochondrial respiratory chain) to a highly glycolytic metabolism, producing large amounts of lactate (Warburg et al. 1924, 1926). As a result, respiration would become secondary and the mitochondria status in tumor tissues would be somewhat irrelevant for tumor development. Recently, however, mutations in three of the four genes encoding respiratory chain complex II (succinate dehydrogenase, SDH) have been shown to cause paragangliomas (PGLs) or pheochromocytomas. PGLs are neuroendocrine tumors that may secrete catecholamines (Favier et al. 2005), which occur most frequently in the head, neck, adrenal medulla and extra-adrenal sympathetic ganglia. The hereditary form of PGLs, about 30% of cases, is usually characterized by an early onset and a more severe presentation than the sporadic form. In 2000, linkage analysis and positional cloning allowed Baysal et al. (2000) to report the first deleterious mutations in the SDHD gene. Subsequently, a candidate gene approach has led to the identification of mutations in SDHC and SDHB (Astuti et al. 2001; Niemann and Muller 2000). This fueled a strong debate on the mechanism linking SDH deficiency to tumor formation, initial observations suggesting that superoxide overproduction might be at the origin of increased cell proliferation (Rustin 2002). Indeed, the mev1 mutant of the worm *Caenorhabditis elegans*, defective in the cytochrome *b* subunit of CII, was found to have a reduced life span ascribed to overproduction of superoxides (Senoo-Matsuda et al. 2001). However, no hyperplasia or indications of abnormal cell proliferation were reported in the *mev1* mutant at variance with other *C. elegans* mutants for proteins which are prone to trigger tumorigenesis when mutated in the homologous protein in mammals, e.g., *cul1* mutant (Piva et al. 2002).

Soon after the discovery that SDH mutations can result in tumor formation, it was shown that these tumors were highly vascularized concomitantly with HIF stabilization and activation of the hypoxia pathway (Gimenez-Roqueplo et al. 2001). As a rule, under normoxic conditions, the HIF-α subunit is continuously ubiquitinated and subsequently degraded by the proteasome (Hickey and Simon 2006). The process of ubiquitination is started by their recognition by the von Hippel–Lindau (VHL) protein, which requires the hydroxylation of two proline residues on HIF-α (Kaelin 2005). The very first step of HIF-α degradation is dependent on this hydroxylation, which is catalyzed by HIF prolyl hydroxylases (PHDs). PHDs belong to the superfamily of the Fe(II)-dependant oxygenases and require reduced iron as a cofactor, α-KG and oxygen as co-substrates, with carbon dioxide and succinate being the products of the reaction (Lee et al. 2004). Under hypoxic conditions, the absence of oxygen prohibits PHD activity, and HIF-α is thus stabilized, allowing for its nuclear translocation and the subsequent activation of the target genes. The involvement of HIFs has been observed in numerous types of tumors, playing an active role in the progression of neoplasia (Gordan and Simon 2007). To make a long story short, it was established that a high intracellular succinate concentration, as measured in SDH-deficient cells and tumors, was responsible for the blockade of PHD activity with the consequent stabilization of the HIF1α protein (Briere et al. 2005a; Selak et al. 2005) (Fig. 2). Conversely, the addition of α-ketoglutarate, the substrate of the PHD, was shown to abolish the nuclear translocation of HIF1α in SDH-defective cells (Briere et al. 2005a). The abnormal organic acid balance thus provided a proficient mechanism linking SDH-deficiency to tumor formation.

Additional support in favor of this latter hypothesis came from the observation that a fumarase defect can lead to leiomyomatosis and renal cell cancer (HLRCC) syndrome (Tomlinson et al. 2002). In this latter case, fumarate, accumulated because of fumarase inactivation, was found to act as a competitive inhibitor of the PHD, thus inducing the abnormal stabilization of HIF-1α. Other structurally related organic acids can also inhibit PHD (MacKenzie et al. 2007). Therefore, a TCA cycle blockade may result in the induction of angiogenesis and tuning up glycolysis during tumorigenesis may be at the origin of the Warburg effect, rather than a blockade of the electron flow (potentially associated with all subtypes of RC defects) (Briere et al. 2005b). It would therefore be at least rather imprudent to invoke SDH mutations as general proof that a RC defect results in tumor formation.

To date, there is no strong evidence that a perturbation of the electron flow through the RC is sufficient to increase cell proliferation and

Mitochondria and Cancer

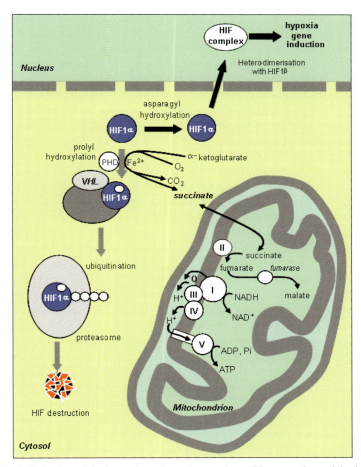

Fig. 2. Induction of the hypoxia pathway by succinate. Upon succinate dehydrogenase (complex II, *II*) blockade, succinate accumulates and is exported from mitochondria to the cytosol. There, it inhibits the prolyl hydroxylase (*PHD*), thus triggering Hif1α stabilization. The nuclear translocation of this latter factor induces an increased transcription of the hypoxia pathway components. *VHL*, Von Hippel–Lindau; *I–V*, the various respiratory chain complexes

constitutes the primary cause of tumor formation: on one hand, none of the nuclear genes encoding RC components or involved the building or maintenance of the chain has been demonstrated to be a tumor suppressor, with the exception of genes encoding RC complex II (or fumarase; see above) (Kroemer 2006). On the other hand, RC-specific poisons are not known as carcinogenic. Finally, patients harboring deleterious mutation in genes encoding RC components are not known to be particularly at risk for tumor formations. This is even true for mutations affecting RC proteins that result in high levels of superoxide production, such as ATPase (complex V) components (Geromel et al. 2001). Even more puzzling is the fact that in a subset of tumor cells (liver and pancreatic tumors), enhanced glycolysis might well require an active production of ATP by the RC. Using both ATP and glucose to produce ADP and glucose-6-phosphate, the hexokinase specifically expressed in most tumor cells is characterized by its poor affinity (high Km, about 10 mM) for glucose, similar to the hexokinase IV (glucokinase) found in normal hepatocytes (Brandon et al. 2006). It possibly allows for an additional capacity for glucose uptake from plasma and increased glucose phosphorylation, by displacement of the cell glucose equilibrium. Interestingly, while the isoform with a low affinity is expressed in most non-tumor cells, in the malignant hepatocellular carcinoma cells and in transformed pancreatic cells, it is largely replaced by a high affinity form (hexokinase II, HKII). This latter form can readily bind VDAC at the outer mitochondrial membrane and utilize the mitochondrial ATP to produce glucose-6-phosphate, thus favoring aerobic glycolysis (Bustamante and Pedersen 1977) as long as the electron transfer chain is working (Fig. 3). In this case, carcinogenesis would in fact require the preservation of the respiratory chain function, at least at a minimal level.

The possibility nevertheless exists that low oxygenation in tumors may secondarily affect mitochondrial function, thus favoring the formation of superoxides and peroxides by the RC. In principle, activated oxygen species might in turn signal both oncogene growth factors and their tyrosine kinase receptors, thus driving cell transformation (Aslan and Ozben 2003), simultaneously promoting HIF1α stabilization by inhibiting the prolyl hydroxylase. The suggested role of superoxides in triggering tumorigenesis has long been advocated in support to a therapeutic use of antioxidants (Nishikawa and Hashida 2006). However,

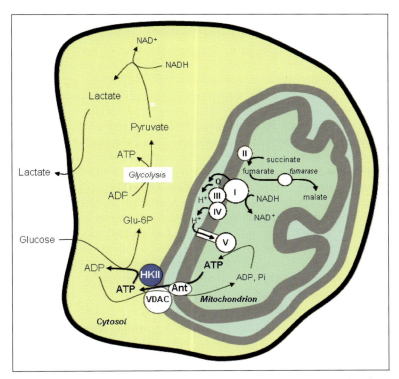

Fig. 3. Interaction between mitochondria and type II hexokinase. The channeling of mitochondrial ATP by type II hexokinase favors the production of glucose 6-phosphate (*G 6-P*) and ultimately glycolysis. *Ant,* adenylate carrier; *HKII,* type II hexokinase; *I–V,* the various complexes of the respiratory chain; *Q,* ubiquinone; *VDAC,* voltage-dependent anion channel

contrasting results from in vitro and in vivo experiments have raised some doubt about the ability of antioxidant enzymes (Lu et al. 1997; Welsh et al. 2002) or molecules to actually fight cancer by such a mechanism. Indeed, one should be aware that most antioxidant molecules also act as prooxidants, possibly accounting for a potential antitumoral activity. For instance, an antioxidant molecule such as melatonin exercises its antiproliferative effect on the growth of rat pituitary prolactin-secreting tumor cells in vitro by damaging mitochondria rather than by quenching superoxides (Yang et al. 2007). Thus, even if increased superoxide production can be evidenced in a subset of cancer cells, we need more evidence to establish that, as a general rule, these superoxides are instrumental in triggering tumorigenesis.

While it is not clear that a defective respiratory chain actually favors tumor formation, mitochondria might instead represent the Achilles tendon of cancer cells. As discussed above, mitochondria house several proapoptotic factors that are simultaneously components of (or closely associated with) the electron transfer chain. Targeting tumors with reagents susceptible to inducing the release of these components has become a fashionable idea (Galluzzi et al. 2006). Thus, cisplatin, one of the most important chemotherapeutic agents ever developed, has been shown to readily interact with mitochondria to trigger apoptosis (Cepeda et al. 2007). Resveratrol, a natural polyphenolic antioxidant, has been reported to possess a cancer chemopreventive potential that has been ascribed to its ability to trigger mitochondrial dysfunction and apoptosis (Fulda and Debatin 2006), mention yet another example.

5 Hypothetical Links Between Apoptosis Resistance and Anaerobic Glycolysis at the Mitochondrial Membrane

As summarized above, mitochondria from cancer cells are relatively resistant against MMP induction, thereby reducing the propensity to undergo apoptosis. In addition, tumor mitochondria are, to some extent, perturbed in their metabolism, often exhibiting reduced OXPHOS. Are these two phenomena mechanistically linked? Unfortunately, there is no simple answer to this question, because there may be multiple links, none of which is firmly established to contribute to oncogenesis or tumor progression.

A first explanation for simultaneous apoptosis inhibition and OXPHOS defects of cancer cells may reside in the composition of mitochondrial membranes. For example, Bcl-2 and Bcl-XL are prominent MMP inhibitors, yet also have direct effects on ATP synthesis in which they act as allosteric activators of ANT (Belzacq et al. 2003). Report-

Mitochondria and Cancer

edly, Bcl-2 and Bcl-XL can also inhibit the capacity of VDAC to exchange metabolites on the outer mitochondrial membrane, an effect that would reduce respiration (Tsujimoto and Shimizu 2007). A functional and structural Bcl-2 homolog, vMIA (for viral mitochondrial inhibitor of apoptosis), which is encoded by cytomegalovirus, acts as a strong inhibitor of apoptosis (via its capacity to inhibit Bax), yet is also an inhibitor of the phosphate carrier, one of the proteins of the ATP synthasome (Poncet et al. 2006). This implies that vMIA reduces ATP generation by OXPHOS, an effect that accounts for the cytopathic effect of cytomegalovirus. No such inhibitory effect was, however, found for Bcl-2 (P. Rustin and G. Kroemer, unpublished data).

Unfortunately, there are no systematic studies on the composition of the outer mitochondrial membrane of cancer cells. However, differences in the composition of the PTPC have been reported, and whether alterations in the abundance of VDAC or ANT isoforms account for dual apoptosis/OXPHOS defects of tumor cells remains to be investigated in detail. One PTPC component, hexokinase, has been shown to associate more vigorously with VDAC in tumor cells than in normal control cells (Pedersen 2007). When associated with VDAC, hexokinase may efficiently couple residual OXPHOS to the initial, rate-limiting step of glycolysis, and simultaneous inhibit MMP, presumably through an effect on the PTPC.

Other links between OXPHOS defects and inhibited apoptosis maybe more indirect. A hyperpolarization of the inner mitochondrial transmembrane potential, as is frequently seen in cancer cells (perhaps secondary to defects in the F1F0 ATPase), can intrinsically reduce the propensity of PTPC opening (Zoratti and Szabo 1995). Total inhibition of the respiratory chain inhibits the activation of the pro-apoptotic Bcl-2 proteins Bax and Bak (Tomiyama et al. 2006). In addition, OXPHOS defects (and in particular mtDNA mutations) might increase the production of ROS and hence activate, via HIF, a transcriptional program that reduces the propensity of the cells to succumb to stress-induced MMP. Major defects in respiratory chain complexes reduce electron flow on the inner mitochondrial membrane and reduce the capacity of certain xenobiotics to elicit ROS generation in mitochondria, thereby abolishing their pro-apoptotic effects. This latter mechanism may explain the fact that ρ° cells (cells that lack mitochondrial DNA and hence OX-

PHOS) are resistant against a series of compounds that induce apoptosis by provoking futile redox cycles in mitochondria (Galluzzi et al. 2006; Kroemer et al. 2007).

Altogether, these examples illustrate the possible links between apoptosis resistance and anaerobic glycolysis in cancer. Future studies will elucidate which among these links, if any, has a preponderant impact on oncogenesis or tumor progression. Similarly, future work will determine whether dual therapeutic interventions that might simultaneously restore apoptosis and target the metabolic alterations linked to cancer might provide valid tools for our combat against cancer.

Acknowledgements. G. Kroemer is supported by Ligue Nationale Contre le Cancer (équipe labellisée), the European Union (Active p53, ChemoRes, Death-Train, TransDeath, RIGHT), Cancéropôle Ile-de-France, Institut National du Cancer, and the Agence Nationale pour la Recherche. P. Rustin is supported by Association Contre les Maladies Mitochondriales et Association Française Contre les Myopathies, Leducq Foundation (CarDiaNet), and the European Union (Eumitocombat).

References

Adams JM, Cory S (2007a) The Bcl-2 apoptotic switch in cancer development and therapy. Oncogene 26:1324–1337

Adams JM, Cory S (2007b) Bcl-2-regulated apoptosis: mechanism and therapeutic potential. Curr Opin Immunol 19:488–496

Aslan M, Ozben T (2003) Oxidants in receptor tyrosine kinase signal transduction pathways. Antioxid Redox Signal 5:781–788

Astuti D, Douglas F, Lennard TW, Aligianis IA, Woodward ER, Evans DG, Eng C, Latif F, Maher ER (2001) Germline SDHD mutation in familial phaeochromocytoma. Lancet 357:1181–1182

Baysal BE, Ferrell RE, Willett-Brozick JE, Lawrence EC, Myssiorek D, Bosch A, van der Mey A, Taschner PE, Rubinstein WS, Myers EN et al (2000) Mutations in SDHD, a mitochondrial complex II gene, in hereditary paraganglioma. Science 287:848–851

Belzacq AS, Vieira HL, Verrier F, Vandecasteele G, Cohen I, Prevost MC, Larquet E, Pariselli F, Petit PX, Kahn A et al (2003) Bcl-2 and Bax modulate adenine nucleotide translocase activity. Cancer Res 63:541–546

Brandon M, Baldi P, Wallace DC (2006) Mitochondrial mutations in cancer. Oncogene 25:4647–4662

Mitochondria and Cancer

17

Brenner C, Kroemer G (2000) Apoptosis. Mitochondria – the death signal integrators. Science 289:1150–1151

Briere JJ, Favier J, Benit P, El Ghouzzi V, Lorenzato A, Rabier D, Di Renzo MF, Gimenez-Roqueplo AP, Rustin P (2005a) Mitochondrial succinate is instrumental for HIF1alpha nuclear translocation in SDHA-mutant fibroblasts under normoxic conditions. Hum Mol Genet 14:3263–3269

Briere JJ, Favier J, Ghouzzi VE, Djouadi F, Benit P, Gimenez AP, Rustin P (2005b) Succinate dehydrogenase deficiency in human. Cell Mol Life Sci 62:2117–2314

Bustamante E, Pedersen PL (1977) High aerobic glycolysis of rat hepatoma cells in culture: role of mitochondrial hexokinase. Proc Natl Acad Sci U S A 74:3735–3739

Cepeda V, Fuertes MA, Castilla J, Alonso C, Quevedo C, Perez JM (2007) Biochemical mechanisms of cisplatin cytotoxicity. Anticancer Agents Med Chem 7:3–18

Costantini P, Jacotot E, Decaudin D, Kroemer G (2000) Mitochondrion as a novel target of anticancer chemotherapy. J Natl Cancer Inst 92:1042–1053

Decaudin D, Castedo M, Nemati F, Beurdeley-Thomas A, De Pinieux G, Caron A, Pouillart P, Wijdenes J, Rouillard D, Kroemer G, Poupon MF (2002) Peripheral benzodiazepine receptor ligands reverse apoptosis resistance of cancer cells in vitro and in vivo. Cancer Res 62:1388–1393

Deng J, Carlson N, Takeyama K, Dal Cin P, Shipp M, Letai A (2007) BH3 profiling identifies three distinct classes of apoptotic blocks to predict response to ABT-737 and conventional chemotherapeutic agents. Cancer Cell 12:171–185

Favier J, Briere JJ, Strompf L, Amar L, Filali M, Jeunemaitre X, Rustin P, Gimenez-Roqueplo AP (2005) Hereditary paraganglioma/pheochromocytoma and inherited succinate dehydrogenase deficiency. Horm Res 63:171–179

Fulda S, Debatin KM (2006) Resveratrol modulation of signal transduction in apoptosis and cell survival: a mini-review. Cancer Detect Prev 30:217–223

Galluzzi L, Larochette N, Zamzami N, Kroemer G (2006) Mitochondria as therapeutic targets for cancer chemotherapy. Oncogene 25:4812–4830

Galluzzi L, Maiuri MC, Vitale I, Zischka H, Castedo M, Zitvogel L, Kroemer G (2007) Cell death modalities: classification and pathophysiological implications. Cell Death Differ 14:1237–1243

Geromel V, Kadhom N, Cebalos-Picot I, Ouari O, Polidori A, Munnich A, Rotig A, Rustin P (2001) Superoxide-induced massive apoptosis in cultured skin fibroblasts harboring the neurogenic ataxia retinitis pigmentosa (NARP) mutation in the ATPase-6 gene of the mitochondrial DNA. Hum Mol Genet 10:1221–1228

Gimenez-Roqueplo AP, Favier J, Rustin P, Mourad JJ, Plouin PF, Corvol P, Rotig A, Jeunemaitre X (2001) The R22X mutation of the SDHD gene in hereditary paraganglioma abolishes the enzymatic activity of complex II in the mitochondrial respiratory chain and activates the hypoxia pathway. Am J Hum Genet 69:1186–1197

Gordan JD, Simon MC (2007) Hypoxia-inducible factors: central regulators of the tumor phenotype. Curr Opin Genet Dev 17:71–77

Hickey MM, Simon MC (2006) Regulation of angiogenesis by hypoxia and hypoxia-inducible factors. Curr Top Dev Biol 76:217–257

Jiang X, Wang X (2004) Cytochrome C-mediated apoptosis. Annu Rev Biochem 73:87–106

Kaelin WG (2005) Proline hydroxylation and gene expression. Annu Rev Biochem 74:115–128

Kluck RM, Bossy-Wetzel E, Green DR, Newmeyer DD (1997) The release of cytochrome c from mitochondria: a primary site for Bcl-2 regulation of apoptosis. Science 275:1132–1136

Kroemer G (2006) Mitochondria in cancer. Oncogene 25:4630–4632

Kroemer G, Petit P, Zamzami N, Vayssiere JL, Mignotte B (1995) The biochemistry of programmed cell death. FASEB J 9:1277–1287

Kroemer G, El-Deiry WS, Golstein P, Peter ME, Vaux D, Vandenabeele P, Zhivotovsky B, Blagosklonny MV, Malorni W, Knight RA et al (2005) Classification of cell death: recommendations of the Nomenclature Committee on Cell Death. Cell Death Differ 12 [Suppl 2]:1463–1467

Kroemer G, Galluzzi L, Brenner C (2007) Mitochondrial membrane permeabilization in cell death. Physiol Rev 87:99–163

Le Bras M, Borgne-Sanchez A, Touat Z, El Dein OS, Deniaud A, Maillier E, Lecellier G, Rebouillat D, Lemaire C, Kroemer G et al (2006) Chemosensitization by knockdown of adenine nucleotide translocase-2. Cancer Res 66:9143–9152

Lee JW, Bae SH, Jeong JW, Kim SH, Kim KW (2004) Hypoxia-inducible factor (HIF-1) alpha: its protein stability and biological functions. Exp Mol Med 36:1–12

Li P, Nijhawan D, Budihardjo I, Srinivasula SM, Ahmad M, Alnemri ES, Wang X (1997) Cytochrome c and dATP-dependent formation of Apaf-1/caspase-9 complex initiates an apoptotic protease cascade. Cell 91:479–489

Liu X, Kim CN, Yang J, Jemmerson R, Wang X (1996) Induction of apoptotic program in cell-free extracts: requirement for dATP and cytochrome c. Cell 86:147–157

Lu YP, Lou YR, Yen P, Newmark HL, Mirochnitchenko OI, Inouye M, Huang MT (1997) Enhanced skin carcinogenesis in transgenic mice with high expression of glutathione peroxidase or both glutathione peroxidase and superoxide dismutase. Cancer Res 57:1468–1474

MacKenzie ED, Selak MA, Tennant DA, Payne LJ, Crosby S, Frederiksen CM, Watson DG, Gottlieb E (2007) Cell-permeating alpha-ketoglutarate derivatives alleviate pseudohypoxia in succinate dehydrogenase-deficient cells. Mol Cell Biol 27:3282–3289

Niemann S, Muller U (2000) Mutations in SDHC cause autosomal dominant paraganglioma, type 3. Nat Genet 26:268–270

Nishikawa M, Hashida M (2006) Inhibition of tumour metastasis by targeted delivery of antioxidant enzymes. Expert Opin Drug Deliv 3:355–369

Oltersdorf T, Elmore SW, Shoemaker AR, Armstrong RC, Augeri DJ, Belli BA, Bruncko M, Deckwerth TL, Dinges J, Hajduk PJ et al (2005) An inhibitor of Bcl-2 family proteins induces regression of solid tumours. Nature 435:677–681

Pedersen PL (2007) Warburg, me and hexokinase 2: multiple discoveries of key molecular events underlying one of cancers' most common phenotypes, the "Warburg Effect", i.e., elevated glycolysis in the presence of oxygen. J Bioenerg Biomembr 39:211–222

Piva R, Liu J, Chiarle R, Podda A, Pagano M, Inghirami G (2002) In vivo interference with Skp1 function leads to genetic instability and neoplastic transformation. Mol Cell Biol 22:8375–8387

Poncet D, Pauleau AL, Szabadkai G, Vozza A, Scholz SR, Le Bras M, Briere JJ, Jalil A, Le Moigne R, Brenner C et al (2006) Cytopathic effects of the cytomegalovirus-encoded apoptosis inhibitory protein vMIA. J Cell Biol 174:985–996

Rustin P (2002) Mitochondria, from cell death to proliferation. Nat Genet 30:352–353

Selak MA, Armour SM, MacKenzie ED, Boulahbel H, Watson DG, Mansfield KD, Pan Y, Simon MC, Thompson CB, Gottlieb E (2005) Succinate links TCA cycle dysfunction to oncogenesis by inhibiting HIF-alpha prolyl hydroxylase. Cancer Cell 7:77–85

Senoo-Matsuda N, Yasuda K, Tsuda M, Ohkubo T, Yoshimura S, Nakazawa H, Hartman PS, Ishii N (2001) A defect in the cytochrome b large subunit in complex II causes both superoxide anion overproduction and abnormal energy metabolism in *Caenorhabditis elegans*. J Biol Chem 276:41553–41558

Susin SA, Zamzami N, Castedo M, Hirsch T, Marchetti P, Macho A, Daugas E, Geuskens M, Kroemer G (1996) Bcl-2 inhibits the mitochondrial release of an apoptogenic protease. J Exp Med 184:1331–1341

Susin SA, Zamzami N, Castedo M, Daugas E, Wang HG, Geley S, Fassy F, Reed JC, Kroemer G (1997) The central executioner of apoptosis: multiple connections between protease activation and mitochondria in Fas/APO-1/CD95- and ceramide-induced apoptosis. J Exp Med 186:25–37

Susin SA, Lorenzo HK, Zamzami N, Marzo I, Snow BE, Brothers GM, Mangion J, Jacotot E, Costantini P, Loeffler M et al (1999) Molecular characterization of mitochondrial apoptosis-inducing factor. Nature 397:441–446

Tomiyama A, Serizawa S, Tachibana K, Sakurada K, Samejima H, Kuchino Y, Kitanaka C (2006) Critical role for mitochondrial oxidative phosphorylation in the activation of tumor suppressors Bax and Bak. J Natl Cancer Inst 98:1462–1473

Tomlinson IP, Alam NA, Rowan AJ, Barclay E, Jaeger EE, Kelsell D, Leigh I, Gorman P, Lamlum H, Rahman S et al (2002) Germline mutations in FH predispose to dominantly inherited uterine fibroids, skin leiomyomata and papillary renal cell cancer. Nat Genet 30:406–410

Tsujimoto Y, Shimizu S (2007) Role of the mitochondrial membrane permeability transition in cell death. Apoptosis 12:835–840

Warburg O, Poesener K, Negelein E (1924) Über den Stoffwechsel der Tumoren [On metabolism of tumors]. Biochem Z 152:319–344

Warburg O, Wind F, Negelein E (1926) The metabolism of tumors in the body. J Gen Physiol 8:519

Welsh SJ, Bellamy WT, Briehl MM, Powis G (2002) The redox protein thioredoxin-1 (Trx-1) increases hypoxia-inducible factor 1alpha protein expression: Trx-1 overexpression results in increased vascular endothelial growth factor production and enhanced tumor angiogenesis. Cancer Res 62:5089–5095

Yagoda N, von Rechenberg M, Zaganjor E, Bauer AJ, Yang WS, Fridman DJ, Wolpaw AJ, Smukste I, Peltier JM, Boniface JJ et al (2007) RAS-RAF-MEK-dependent oxidative cell death involving voltage-dependent anion channels. Nature 447:864–868

Yang J, Liu X, Bhalla K, Kim CN, Ibrado AM, Cai J, Peng TI, Jones DP, Wang X (1997) Prevention of apoptosis by Bcl-2: release of cytochrome c from mitochondria blocked. Science 275:1129–1132

Yang QH, Xu JN, Xu RK, Pang SF (2007) Antiproliferative effects of melatonin on the growth of rat pituitary prolactin-secreting tumor cells in vitro. J Pineal Res 42:172–179

Zamzami N, Kroemer G (2001) The mitochondrion in apoptosis: how Pandora's box opens. Nat Rev Mol Cell Biol 2:67–71

Zamzami N, Susin SA, Marchetti P, Hirsch T, Gomez-Monterrey I, Castedo M, Kroemer G (1996) Mitochondrial control of nuclear apoptosis. J Exp Med 183:1533–1544

Mitochondria and Cancer

Zoratti M, Szabo I (1995) The mitochondrial permeability transition. Biochim Biophys Acta 1241:139–176

Zou H, Henzel WJ, Liu X, Lutschg A, Wang X (1997) Apaf-1, a human protein homologous to C elegans CED-4, participates in cytochrome c-dependent activation of caspase-3. Cell 90:405–413

Ernst Schering Foundation Symposium Proceedings, Vol. 4, pp. 23–34
DOI 10.1007/2789_2008_087
© Springer-Verlag Berlin Heidelberg
Published Online: 06 June 2008

Role of the Metabolic Stress Responses of Apoptosis and Autophagy in Tumor Suppression

E. White[(✉)]

Department of Molecular Biology and Biochemistry, Rutgers University, Cancer Institute of New Jersey, CABM-Room 140, 679 Hoes Lane, 08854 Piscataway, USA
email: *ewhite@cabm.rutgers.edu*

1	Origins of Metabolic Stress in Tumors	24
2	Activation of Tumor Suppression by Apoptosis in Response to Cellular Stress	25
3	Modulation of the Apoptotic Response by Cancer Therapy	25
4	Inactivation of Apoptosis in Tumor Progression	26
5	Autophagy Mediates Tumor Cell Survival to Metabolic Stress	27
6	Autophagy Is a Tumor Suppression Mechanism	29
7	Autophagy Suppresses Cell Death and Inflammation to Limit Tumor Progression	30
8	Autophagy Prevents Genome Damage to Suppress Tumorigenesis	31
9	Modulation of Tumor Cell Metabolism in Cancer Therapy	32
References		32

Abstract. Metabolic stress is an important stimulus that promotes apoptosis-mediated tumor suppression. Metabolic stress arises in tumors from multiple factors that include insufficient nutrient supply caused by deficient angiogenesis and high metabolic demand of unrestrained cell proliferation. The high metabolic demand of tumor cells is only exacerbated by reliance on the inefficient process of glycolysis for energy production. Recently it has become clear

24   E. White

that tumor cells survive metabolic stress through the catabolic process of autophagy. Autophagy also functions as a tumor suppression mechanism by preventing cell death and inflammation and by protecting the genome from damage and genetic instability. How autophagy protects the genome is not yet clear but may be related to its roles in sustaining metabolism or in the clearance of damaged proteins and organelles and the mitigation of oxidative stress. These findings illuminate the important role of metabolism in cancer progression and provide specific predictions for metabolic modulation in cancer therapy.

1   Origins of Metabolic Stress in Tumors

Metabolic stress is a common occurrence in human tumors and is caused by multiple factors. Tumors initially lack a blood supply and when growth progresses towards a mass in the range of 1 mm in diameter, passive diffusion of nutrients is insufficient to sustain those tumor cells in the center of the mass, which suffer metabolic stress (Folkman 2006). Continuation of tumor growth requires angiogenesis to supply nutrients, and recruitment of a blood supply to the tumor can ameliorate metabolic stress. Using hypoxia as a marker for metabolic stress in tumors, induction of metabolic stress during tumor formation and stress abatement following angiogenesis is readily apparent (Nelson et al. 2004). Following successful tumor angiogenesis, metabolic stress is still a factor since blood vessel formation is abnormal and subject to intermittent collapse that inflicts metabolic stress in established tumors. Thus, unlike normal tissues, tumors lack a constant, uninterrupted nutrient supply and thereby are regularly assaulted by bouts of metabolic stress.

Another contributing factor to metabolic stress in tumors is their high metabolic demand caused by unregulated, unrelenting tumor cell growth, a hallmark of cancer (Hanahan and Weinberg 2000). Normal tissues have strict controls that limit cell division to specific developmental periods and circumstances that are tightly linked to nutrient and growth factor availability. Tumor cells lack these controls and proliferate independently, by either autocrine mechanisms or despite the absence of growth factors and nutrients. This dissociation of growth control from nutrient availability is a significant contributor to metabolic stress in tumors that leads to cellular damage and cell death by metabolic catastrophe (Jin et al. 2007). Finally, inefficient energy production in

Role of the Metabolic Stress in Cancer

tumor cells due to metabolism dominated by glycolysis has long been recognized as critical distinction between normal and tumor cells (Warburg 1956). The juxtaposition of restricted nutrient supply, high energy demand, and inefficient energy production exemplifies the fundamental importance of metabolic stress to tumor physiology.

2 Activation of Tumor Suppression by Apoptosis in Response to Cellular Stress

Cellular stress is a well-known trigger of cell death by apoptosis, which is a critical tumor suppression mechanism. Conceptually, cellular stress that results in excessive cell damage activates apoptosis as a means to eliminate dysfunctional and potentially dangerous cells that can acquire mutations and progress toward cancer. There are multiple independent pathways by which apoptosis is triggered in tumors (Adams and Cory 2007; Gelinas and White 2005). Two stress-related pathways that signal apoptosis are the DNA damage and cellular stress response pathway controlled by the p53 tumor suppressor, and the metabolic stress pathway (Fig. 1) (Jin et al. 2007; Jin and White 2007; White 2006).

p53 activates apoptosis in part through transcriptional upregulation of the pro-apoptotic BH3-only proteins Puma and Noxa (Vousden and Lane 2007), whereas metabolic stress signals through the pro-apoptotic BH3-only protein Bim (Nelson et al. 2004; Tan et al. 2005). BH3-only proteins, in turn, signal apoptosis through the multidomain proapoptotic Bax and Bak proteins, which permeabilize the outer mitochondrial membrane causing the release of factors that promote caspase activation leading to cell death (Adams and Cory 2007). How Bim is activated in response to metabolic stress to induce apoptosis is not known, but Bim is a critical epithelial tumor suppressor in the p53-independent apoptotic pathway (Degenhardt et al. 2002; Tan et al. 2005).

3 Modulation of the Apoptotic Response by Cancer Therapy

One of the advantages of a comprehensive knowledge of apoptosis regulation is the ability to use that information to rationally tailor cancer

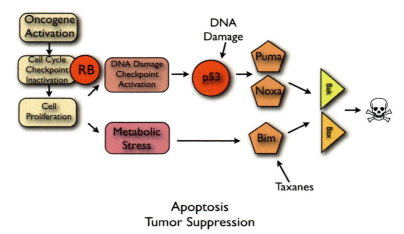

Fig. 1. Activation of the apoptotic tumor suppression response to metabolic stress. See text for explanation

therapies aimed at activating or enhancing the apoptotic response. As such, the development of cancer treatments specifically targeting the apoptotic pathway shows great promise (Fesik 2005; Wu et al. 2007). For example, promoting p53 activation by inducing DNA damage with cytotoxic chemotherapy is one means to shift tumor cells toward apoptosis during cancer therapy (Fig. 1). Another example is the induction of Bim by the taxane class of chemotherapeutic drugs, which results in Bim-mediated apoptosis (Fig. 1) and tumor regression in preclinical models (Tan et al. 2005). These types of therapeutic strategies can shift the balance in tumor cells from survival to death to facilitate treatment response. Indeed, many of the newer noncytotoxic targeted therapies that inhibit signal transduction, relieve apoptosis suppression, thereby favoring tumor cell death.

4 Inactivation of Apoptosis in Tumor Progression

Cancer cells, however, can defeat proapoptotic tumor suppression mechanisms by inactivating apoptosis in various ways that enable tumor progression and treatment resistance (Fig. 2) (Adams and Cory 2007). For example, anti-apoptotic Bcl-2 family members are overexpressed in many tumors, bind and inhibit Bax and Bak to block apoptosis, enabling tumor cell survival, tumor progression and treatment resistance. Analogously, many viruses have acquired these same Bcl-2-like anti-apoptotic mechanisms among their repertoire of oncogenes (Cuconati and White 2002). One example is the E1B19K gene product of the DNA tumor virus adenovirus (Fig. 2), which blocks apoptosis by binding and inhibiting Bax and Bak to facilitate oncogenesis (Cuconati and White 2002; Nelson et al. 2004; White 2006). Tumors acquire mechanisms for inactivating apoptosis that can also be indirect. One example is the upregulation of the ubiquitin ligase for p53, Mdm-2, which promotes p53 degradation in the ubiquitin proteasome pathway to promote tumorigenesis (Vousden and Lane 2007). Another is the phosphorylation and proteasome-mediated degradation of Bim by the Map kinase Erk, which inactivates apoptosis, thereby promoting tumorigenesis and resistance to taxane chemotherapy (Fig. 2) (Tan et al. 2005). Defining how tumors defeat the apoptotic response as exemplified above can dictate rational approaches to anti-cancer therapies.

5 Autophagy Mediates Tumor Cell Survival to Metabolic Stress

Once tumor cells evolve defects in apoptosis, they fail to die in response to metabolic stress, but this does not suffice to explain how these undead cells survive long periods of metabolic stress. By simulating in vivo metabolic stress in tumors in vitro, we discovered that metabolic stress activates the catabolic survival pathway of autophagy (Degenhardt et al. 2006). This autophagy allows tumor cells to cannibalize themselves to sustain their metabolism during extended periods of starvation (Degenhardt et al. 2006). Furthermore, autophagy in tumors in vivo localizes to regions of metabolic stress and tumor cells with defects in autophagy fail to survive in these tumor regions (Fig. 3A) (Degenhardt et al. 2006; Karantza-Wadsworth et al. 2007; Mathew et al. 2007b). These findings revealed that autophagy is an important component of tumor physiology linked to metabolic stress survival.

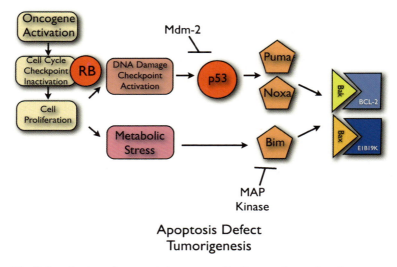

Fig. 2. Inactivation of apoptosis in cancer cells. See text for explanation

Starvation promotes autophagy, which through the action of a host of essential autophagy regulators such as Beclin1 and ATG proteins, double membrane vesicles form that engulf cytoplasm and organelles that traffic this cargo to lysosomes for degradation (Klionsky 2007; Levine and Kroemer 2008; Mizushima 2007). By recycling intracellular nutrients, autophagy can support cellular metabolism during nutrient and growth factor deprivation. In mammalian development, autophagy is essential to sustain metabolism to enable survival during the neonatal starvation period (Kuma et al. 2004). Autophagy also plays a critical role in maintaining protein and organelle quality control exemplified by the accumulation of protein aggregates and aberrant mitochondria associated with neuronal degeneration in mice (Hara et al. 2006; Komatsu et al. 2005, 2006, 2007). Thus, autophagy is an important component of the metabolic stress response that promotes normal cell survival by providing an alternate energy source and by degrading damaged proteins and organelles.

A. Autophagy sustains tumor cell survival to metabolic stress

Tumor Growth, Metabolic Stress, Autophagy, Survival   Angiogenesis, Tumor Growth

B. Defects in autophagy promote cell death, inflammation and and tumor growth

Tumor Growth, Metabolic Stress, Chronic Necrosis/Apoptosis, Inflammation, Angiogenesis
Tumor Progression

C. Defects in autophagy promote genome damage and tumor growth

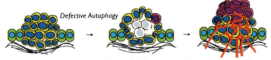

Tumor Growth, Metabolic Stress, DNA damage, Mutation, Tumor Progression

Fig. 3A–C. Dual roles for autophagy in suppressing tumorigenesis. **A** Cells at the center of a tumor are subjected to metabolic stress can either undergo apoptosis or, in this illustration, can activate autophagy to enable long-term survival. Eventually, with acquisition of a blood supply through angiogenesis, tumor progression can occur. **B** Defects in autophagy promote either necrotic or apoptotic cell death and the inflammation and cytokine production that results can promote tumorigenesis by a non-cell-autonomous mechanism. **C** Defects in autophagy cause DNA damage and genome instability due to the failure to mitigate metabolic stress. The accelerated mutation rate that results can facilitate tumor progression by a cell-autonomous mechanism

6   Autophagy Is a Tumor Suppression Mechanism

Although autophagy promotes tumor cell survival, paradoxically, allelic loss of the essential autophagy gene *beclin1* is found with high frequency in human cancers, and *beclin1* [+/–] mutant mice are tumor-prone (Liang et al. 1999; Qu et al. 2003; Yue et al. 2003). Furthermore, allelic loss of *beclin1* and defective autophagy promotes tumorigenesis of immortal mouse kidney and mammary epithelial cells (Degenhardt et al. 2006; Karantza-Wadsworth et al. 2007; Mathew et al. 2007b). It then became a challenge to decipher the mechanism by which loss of function of a survival pathway of autophagy promotes tumor growth. Although on the surface it may appear counterintuitive that loss of a survival function such as autophagy causes tumorigenesis, this is not the first time this situation has occurred. Defects in DNA repair, for example, compromise cell survival to DNA damage, yet the accelerated mutation rate promotes tumorigenesis (Hanahan and Weinberg 2000). In this setting, the elevated mutation rate that results from DNA repair deficiency compensates for the reduced survival to DNA damage by expediting tumor evolution. As such, enhanced tumor cell survival is not always associated with promotion of tumorigenesis.

7 Autophagy Suppresses Cell Death and Inflammation to Limit Tumor Progression

In tumor cells as in normal cells, autophagy promotes survival to metabolic stress. In apoptosis-competent tumor cells, autophagy suppresses apoptosis in response to metabolic stress (Boya et al. 2005; Degenhardt et al. 2006; Karantza-Wadsworth et al. 2007; Mathew et al. 2007b). In apoptosis-defective tumor cells, autophagy enables long-term survival to metabolic stress such that restoration of nutrients permits efficient recovery (Degenhardt et al. 2006; Karantza-Wadsworth et al. 2007; Mathew et al. 2007b). This prosurvival function of autophagy prevents tumor cell death, and as a result, the inflammation, macrophage infiltration, and tumor-promoting cytokine production that result from excessive tumor cell death is prevented (Degenhardt et al. 2006; Jin and White 2007; Karantza-Wadsworth and White 2007; Mathew et al. 2007a; Mathew and White 2007). This chronic inflammation in autophagy-defective tumors is associated with enhanced tumor growth (Fig. 3B,C). In tumors where apoptosis is intact, autophagy defects can promote chronic apoptosis in response to metabolic stress, and excess apoptotic tumor cells can promote macrophage infiltration associated

Role of the Metabolic Stress in Cancer 31

with poor prognosis (Fig. 3B) (Condeelis and Pollard 2006). In tumors with defects in apoptosis, deficient autophagy results in chronic necrotic cell death due to metabolic catastrophe in response to metabolic stress analogous to a corrupted would-healing response associated with poor prognosis (Fig. 3B) (Degenhardt et al. 2006; Jin et al. 2007; Jin and White 2007; Mathew et al. 2007a). While the role of inflammation in tumorigenesis is complex, many aspects of the inflammatory response are associated with tumor progression (Balkwill et al. 2005). Thus, by preserving cell health and viability, autophagy serves as an important non-cell-autonomous tumor-suppression mechanism by limiting inflammation (Degenhardt et al. 2006; Mathew et al. 2007a).

8 Autophagy Prevents Genome Damage to Suppress Tumorigenesis

While autophagy permits tumor cells to tolerate metabolic stress, it serves an important protective function by limiting the accumulation of damaged proteins and organelles (Komatsu et al. 2007; Komatsu et al. 2005), which may ultimately protect the genome and suppresses tumor progression (Mathew et al. 2007a). Tumorigenesis commonly requires stable genetic or epigenetic changes acquired through mutation (Hanahan and Weinberg 2000). Interestingly, accumulation of damaged proteins and organelles can be associated with elevated oxidative stress that may directly or indirectly promote genome damage and mutation. Indeed, autophagy-defective immortal mouse kidney and mammary epithelial cells have an elevated DNA damage response in metabolic stress and are prone to gene amplification and chromosome gains and losses (Karantza-Wadsworth et al. 2007; Mathew et al. 2007b). These findings provide evidence that autophagy protects the genome, providing the first link between metabolism and limiting DNA damage and mutation that can promote tumorigenesis by a cell-autonomous mechanism (Fig. 3C) (Jin and White 2007; Karantza-Wadsworth and White 2007; Mathew et al. 2007a; Mathew and White 2007). Determining the molecular mechanism by which autophagy protects cells from genome damage will be of great interest.

9 Modulation of Tumor Cell Metabolism in Cancer Therapy

The findings described above illustrate that the metabolic stress response is linked to both apoptosis and autophagy, and understanding these pathways is essential for developing cancer therapies. As autophagy is clearly a survival pathway utilized by tumor cells to survive metabolic stress, inhibitors of autophagy may be therapeutically useful (Jin et al. 2007; Jin and White 2007; Karantza-Wadsworth and White 2007; Mathew et al. 2007a; Mathew and White 2007). In this setting, it will be important to induce acute rather than chronic tumor cell death to avoid inflammation and genome instability associated with impairment of autophagy. Furthermore, since autophagy provides a protective tumor-suppression function, autophagy stimulators may be particularly useful in the setting of cancer prevention. Therapeutically stimulating autophagy in individuals at risk may suppress cell death, inflammation, and genome damage, thereby restricting tumor progression (Jin et al. 2007; Jin and White 2007; Karantza-Wadsworth and White 2007; Mathew et al. 2007a; Mathew and White 2007). Finally, as many tumors have defects in autophagy in conjunction with high metabolic demand that is reliant on the inefficient process of glycolysis for energy production, strategies to inflict therapeutic starvation should be considered (Jin et al. 2007). In that way, we can take advantage of the most fundamental factor that discriminates normal cells from cancer cells (Warburg 1956).

References

Adams JM, Cory S (2007) Bcl-2-regulated apoptosis: mechanism and therapeutic potential. Curr Opin Immunol 19:488–496

Balkwill F, Charles KA, Mantovani A (2005) Smoldering and polarized inflammation in the initiation and promotion of malignant disease. Cancer Cell 7:211–217

Boya P, Gonzalez-Polo RA, Casares N, Perfettini JL, Dessen P, Larochette N, Metivier D, Meley D, Souquere S, Yoshimori T et al (2005) Inhibition of macroautophagy triggers apoptosis. Mol Cell Biol 25:1025–1040

Condeelis J, Pollard JW (2006) Macrophages: obligate partners for tumor cell migration, invasion and metastasis. Cell 124:263–266

Role of the Metabolic Stress in Cancer 33

Cuconati A, White E (2002) Viral homologs of BCL-2: role of apoptosis in the regulation of virus infection. Genes Dev 16:2465–2478

Degenhardt K, Chen G, Lindsten T, White E (2002) BAX and BAK mediate p53-independent suppression of tumorigenesis. Cancer Cell 2:193–203

Degenhardt K, Mathew R, Beaudoin B, Bray K, Anderson D, Chen G, Mukherjee C, Shi Y, Gelinas C, Fan Y et al (2006) Autophagy promotes tumor cell survival and restricts necrosis, inflammation and tumorigenesis. Cancer Cell 10:51–64

Fesik SW (2005) Promoting apoptosis as a strategy for cancer drug discovery. Nat Rev 5:876–885

Folkman J (2006) Angiogenesis. Annu Rev Med 57:1–18

Gelinas C, White E (2005) BH3-only proteins in control: specificity regulates MCL-1 and BAK-mediated apoptosis. Genes Dev 19:1263–1268

Hanahan D, Weinberg RA (2000) The hallmarks of cancer. Cell 100:57–70

Hara T, Nakamura K, Matsui M, Yamamoto A, Nakahara Y, Suzuki-Migishima R, Yokoyama M, Mishima K, Saito I, Okano H et al (2006) Suppression of basal autophagy in neural cells causes neurodegenerative disease in mice. Nature 441:885–889

Jin S, White E (2007) Role of autophagy in cancer: management of metabolic stress. Autophagy 3:28–31

Jin S, DiPaola RS, Mathew R, White E (2007) Metabolic catastrophe as a means to cancer cell death. J Cell Sci 120:379–383

Karantza-Wadsworth V, White E (2007) Role of autophagy in breast cancer. Autophagy 3:610–613

Karantza-Wadsworth V, Patel S, Kravchuk O, Chen G, Mathew R, Jin S, White E (2007) Autophagy mitigates metabolic stress and genome damage in mammary tumorigenesis. Genes Dev 21:1621–1635

Klionsky DJ (2007) Autophagy: from phenomenology to molecular understanding in less than a decade. Nat Rev Mol Cell Biol 8:931–937

Komatsu M, Waguri S, Ueno T, Iwata J, Murata S, Tanida I, Ezaki J, Mizushima N, Ohsumi Y, Uchiyama Y et al (2005) Impairment of starvation-induced and constitutive autophagy in Atg7-deficient mice. J Cell Biol 169:425–434

Komatsu M, Waguri S, Chiba T, Murata S, Iwata J, Tanida I, Ueno T, Koike M, Uchiyama Y, Kominami E et al (2006) Loss of autophagy in the central nervous system causes neurodegeneration in mice. Nature 441:880–884

Komatsu M, Waguri S, Koike M, Sou YS, Ueno T, Hara T, Mizushima N, Iwata JI, Ezaki J, Murata S et al (2007) Homeostatic levels of p62 control cytoplasmic inclusion body formation in autophagy-deficient mice. Cell 131:1149–1163

Kuma A, Hatano M, Matsui M, Yamamoto A, Nakaya H, Yoshimori T, Ohsumi Y, Tokuhisa T, Mizushima N (2004) The role of autophagy during the early neonatal starvation period. Nature 432:1032–1036

Levine B, Kroemer G (2008) Autophagy in the pathogenesis of disease. Cell 132:27–42

Liang XH, Jackson S, Seaman M, Brown K, Kempkes B, Hibshoosh H, Levine B (1999) Induction of autophagy and inhibition of tumorigenesis by beclin 1. Nature 402:672–676

Mathew R, White E (2007) Why sick cells produce tumors: the protective role of autophagy. Autophagy 3:502–505

Mathew R, Karantza-Wadsworth V, White E (2007a) Role of autophagy in cancer. Nat Rev Cancer 7:961–967

Mathew R, Kongara S, Beaudoin B, Karp CM, Bray K, Degenhardt K, Chen G, Jin S, White E (2007b) Autophagy suppresses tumor progression by limiting chromosomal instability. Genes Dev 21:1367–1381

Mizushima N (2007) Autophagy: process and function. Genes Dev 21:2861–2873

Nelson DA, Tan TT, Rabson AB, Anderson D, Degenhardt K, White E (2004) Hypoxia and defective apoptosis drive genomic instability and tumorigenesis. Genes Dev 18:2095–2107

Qu X, Yu J, Bhagat G, Furuya N, Hibshoosh H, Troxel A, Rosen J, Eskelinen EL, Mizushima N, Ohsumi Y et al (2003) Promotion of tumorigenesis by heterozygous disruption of the beclin 1 autophagy gene. J Clin Invest 112:1809–1820

Tan TT, Degenhardt K, Nelson DA, Beaudoin B, Nieves-Neira W, Bouillet P, Villunger A, Adams JM, White E (2005) Key roles of BIM-driven apoptosis in epithelial tumors and rational chemotherapy. Cancer Cell 7:227–238

Vousden KH, Lane DP (2007) p53 in health and disease. Nat Rev Mol Cell Biol 8:275–283

Warburg O (1956) On respiratory impairment in cancer cells. Science 124:269–270

White E (2006) Mechanisms of apoptosis regulation by viral oncogenes in infection and tumorigenesis. Cell Death Differ 13:1371–1377

Wu H, Tschopp J, Lin SC (2007) Smac mimetics and TNFalpha: a dangerous liaison? Cell 131:655–658

Yue Z, Jin S, Yang C, Levine AJ, Heintz N (2003) Beclin 1, an autophagy gene essential for early embryonic development, is a haploinsufficient tumor suppressor. Proc Natl Acad Sci U S A 100:15077–15082

Ernst Schering Foundation Symposium Proceedings, Vol. 4, pp. 35–53
DOI 10.1007/2789_2008_088
© Springer-Verlag Berlin Heidelberg
Published Online: 06 June 2008

The Interplay Between MYC and HIF in the Warburg Effect

C.V. Dang(✉)

Department of Medicine, Cell Biology, Molecular Biology and Genetics,
Oncology and Pathology, Johns Hopkins University School of Medicine,
Ross Research Building, Room 1032, 720 Rutland Avenue, 21205 Baltimore, USA
email: *cvdang@jhmi.edu*

1	MYC Is a Major Human Oncogene	
	That Encodes a Transcription Factor	36
2	Myc Target Genes in Tumorigenesis	38
3	Myc-E2F Regulatory Axis	
	and DNA Metabolism and Replication	40
4	Effects of Antioxidants on Myc-Mediated Tumorigenesis,	
	Reactive Oxygen Species and the Hypoxia-Inducible Factor	41
5	Collaboration of MYC and HIF in the Warburg Effect	43
References		46

Abstract. c-MYC and the hypoxia-inducible factors (HIFs) are critical factors
for tumorigenesis in a large number of human cancers. While the normal function of MYC involves the induction of cell proliferation and enhancement of
cellular metabolism, the function of HIF, particularly HIF-1, involves adaptation to the hypoxic microenvironment, including activation of anaerobic glycolysis. When MYC-dependent tumors grow, the hypoxic tumor microenvironment elevates the levels of HIF, such that oncogenic MYC and HIF collaborate
to enhance the cancer cell's metabolic needs through increased uptake of glucose and its conversion to lactate. HIF is also able to attenuate mitochondrial
respiration through the induction of pyruvate dehydrogenase kinase 1 (PDK1),
which in part accounts for the Warburg effect that describes the propensity for
cancers to avidly take up glucose and convert it to lactate with the concurrent
decrease in mitochondrial respiration. Target genes that are common to both

HIF and MYC, such as PDK1, LDHA, HK2, and TFRC, are therefore attractive therapeutic targets, because their coordinate induction by HIF and MYC widens the therapeutic window between cancer and normal tissues.

1 MYC Is a Major Human Oncogene That Encodes a Transcription Factor

MYC is frequently altered in human cancer (Adhikary and Eilers 2005; Dang et al. 2006; Jamerson et al. 2004; Liao and Dickson 2000). A non-exhaustive compilation of studies on alterations of MYC in human cancers is available on-line at www.myccancergene.org. From such a compilation and the literature, one can estimate that beyond Burkitt's lymphoma, which is virtually 100% affected by *MYC* translocations, *MYC* expression is increased in a variety of human cancers, including 40% breast, 70% colon, 90% gynecological, 50% liver, 40% melanoma, 40% medulloblastomas, 65% prostate and 33% small-cell lung cancers. It is notable that *MYC* gene amplication, which portends poor prognosis, has been observed in 15% of human breast cancers, 5%–10% lung cancers and approximately 5% of colon cancers (Fig. 1). Furthermore, *MYC* was found activated in colon cancers when the APC tumor suppressor is lost (He et al. 1998). APC dampens the transcription of the *MYC* gene through direct regulation of the *MYC* activator, TCF4. All of these studies together demonstrate the importance of deregulated *MYC* in human cancers.

The tumorigenic role of *MYC* is underscored by the variety of tumors arising from its ectopic expression in different tissues of transgenic mice (Cory et al. 1987; Langdon et al. 1986; Leder et al. 1986; Pelengaris et al. 1999; Shchors et al. 2006). Subsequent to initial studies demonstrating that forced expression of *myc* in lymphoid tissues resulted in lymphoid hyperplasia and lymphomas, virtually all other studies of constitutive or inducible *myc* in tissues from skin to liver resulted in neoplastic transformation of the targeted tissue. These tumors appear after a lag time, which signifies the requirement of additional genetic alterations for tumorigenesis. In the lymphoid model, inactivation of p53 or ARF appears necessary for Myc-mediated lymphomagenesis (Eischen et al. 1999; Zindy et al. 1998). More recently, missense mutations of

Role of MYC in neoplastic transformation

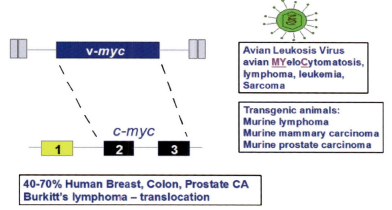

Fig. 1. The MYC proto-oncogene was discovered as the precursor of the retroviral v-myc gene. v-Myc is potently transforming in chickens, giving rise to the myelocytomatosis syndrome, from which the term "MYC" was derived. Deregulation expression of MYC in transgenic animals result in tumorigenesis of the targeted tissue. MYC deregulation also accounts for many human cancers, with the pathognomonic chromosomal translocation and activation of MYC in Burkitt's lymphoma

MYC found in human Burkitt's lymphomas are shown to be defective in triggering apoptosis through Bim, a Bcl-2 antagonist (Hemann et al. 2005). As such, an emerging picture of Myc mediated tumorigenesis includes the concurrent inactivation of apoptotic pathways in some tissues but not others (Pelengaris et al. 1999; Shchors et al. 2006).

The Myc protein was initially localized in cell nuclei, and its function was equivocal, with its role being implicated in DNA replication, RNA splicing, and transcription. Subsequent studies demonstrate the ability of the Myc C-terminal region to oligomerize and that Max is Myc's obligate binding partner (Blackwood and Eisenman 1991; Dang et al. 1989; Prendergast et al. 1991). The heterodimeric Myc-Max protein was found to bind specific core DNA hexameric consensus sequence or E box, 5′-CAC(G/A)TG-3′, as well as related noncanonical

E boxes (Blackwell et al. 1990, 1993; Prendergast and Ziff 1991). In addition, the N-terminal region of Myc, which is required for transformation, fused to the DNA binding domain of the yeast GAL4 transcription, was able to stimulate transcription (Kato et al. 1990). Myc is also capable of transcriptional repression through different mechanisms, including the titration of the Miz-1 transactivator (Adhikary and Eilers 2005; Marhin et al. 1997; Schneider et al. 1997). It is notable that the Myc-Max network is quite complex, since Max also interacts with six other proteins, including Mxd1, Mxd2, Mxd3, Mxd4, Mnt, and Mga (Baudino and Cleveland 2001; Hooker and Hurlin 2006; McArthur et al. 1998). Many of these non-Myc heterodimers could counter the function of Myc. In aggregate, these findings paint an emerging picture of Myc as an oncogenic transcription factor that is involved in a complicated Max-associated network of proteins to regulate gene expression. Recent findings, however, indicate that Myc also has nontranscriptional functions as well as a role in DNA replication. Specifically, the N-terminal domain of Myc stimulates mRNA cap methylation in the absence of the Myc DNA binding domain (Cole and Nikiforov 2006; Cowling and Cole 2007).

2 Myc Target Genes in Tumorigenesis

The biological properties of *MYC* in cell culture and in vivo models suggest that Myc plays a pleiotropic role in the regulation of cell size and proliferation as well as cellular metabolism and adhesion (Adhikary and Eilers 2005; Dang et al. 2006). An accumulating body of literature on Myc target genes is annotated (www.myccancergene.org). To date, direct and indirect Myc target genes have been implicated in a variety of cellular processes (Cole and McMahon 1999; Dang 1999; Fernandez et al. 2003). As one might expect, since Myc induces cell proliferation under specific circumstances, a group of Myc target genes involves cell cycle regulation (Amati et al. 1998; Burgin et al. 1998). Activation of positive cell cycle regulators, such as CDK4, and suppression of negative cell cycle effectors, such as p21, have been reported (Gartel et al. 2001; Hermeking et al. 2000; Mateyak et al. 1999; Wu et al. 2003). In particular, CDK4 has been directly implicated downstream

The Interplay Between MYC and HIF in the Warburg Effect   39

of Myc in the skin tumorigenesis through the used of CDK4 deletion in mice (Miliani de Marval et al. 2004). We reported that the miR-17 cluster, comprising six predicted miRNAs, is a direct Myc target that in part modulates E2F1 translation (O'Donnell et al. 2005). E2F1 in turn directly regulates G1-S transition (Leone et al. 2001; Sears and Nevins 2002). Moreover, another report demonstrates that the miR-17 cluster, which is amplified or overexpressed in human lymphomas and colon cancer, could collaborate with Myc in tumorigenesis in vivo (He et al. 2005). Although increased expression of Myc resulted in increased cell size in lymphocytes and hepatocytes associated with elevation of mRNAs encoding components essential for ribosome biogenesis, which is implicated in *Drosophila* as key for control of cell size, effectors necessary for cell size control have not been clearly delineated in vertebrate systems (Iritani et al. 2002; Kim et al. 2000). In fact, hypomorphs of *myc* in the mouse led to decreased body size associated with fewer cells rather than with smaller cells as found with hypomorphic *dMyc* in *Drosophila* (Gallant et al. 1996; Johnston et al. 1999; Schreiber-Agus et al. 1997; Trumpp et al. 2001). The connection between Myc and ribosomal biogenesis is further strengthened by the recent finding that Myc also directly regulates rRNA synthesis (Arabi et al. 2005; Felton-Edkins et al. 2003; Gomez-Roman et al. 2003; Oskarsson and Trumpp 2005).

Myc has been connected to different metabolic pathways through its initial links to *CAD* (nucleotide metabolism) (Boyd and Farnham 1997), *ODC* (ornithine/spermine metabolism) (Bello-Fernandez et al. 1993), *LDHA* (glucose metabolism) (Shim et al. 1997) and *SHMT2* (single carbon metabolism) (Nikiforov et al. 2002). Global gene expression analysis now connects Myc with diverse metabolic pathways with an overrepresentation of Myc responsive genes involved in glucose and nucleotide metabolism. Although Myc's involvement in regulating glucose metabolism has been well delineated, the connection between Myc and purine and pyrimidine metabolism is not well understood. Our preliminary studies have further connected Myc with the regulation of almost all genes involved in pyrimidine biosynthesis, in addition to *CAD*, and many genes linked to purine biosynthesis involved in the generation of inosine monophosphate (IMP), a precursor of AMP and GMP. Intriguingly, the pathway to produce IMP involves an intermediate, AICAR, which is an activator of AMP kinase (AMPK).

AMPK senses cellular energy status through the levels of AMP (Luo et al. 2005). AICAR, thereby, potentially links nucleotide metabolism to energy regulation since activation of AMP kinase triggers pathways that conserve or generate ATP.

Another theme that emerges is the downregulation of genes involved in cell adhesion by Myc (Dang et al. 2006). This aspect emphasizes the importance of cell-type specificity of Myc's effects, since different cell adhesion molecules are downregulated depending on the cell type in question. One of the intriguing areas that have not been explored is the ability of Myc to regulate other transcriptional factors and how Myc collaborates with other transcription factors in *cis* -regulatory modules to activate or suppress transcription of specific subsets of genes involved in tumorigenesis. Specifically, Myc and the hypoxia-inducible factor (HIF-1) have been implicated in tumor DNA repair and adaptation to the tumor microenvironment (Koshiji et al. 2005). To date, very little is known about modifiers that affect Myc tumorigenic function in vivo. Our preliminary studies also demonstrate a functional link between Myc and E2F1, particularly in the regulation of nucleotide metabolism (Fig. 2).

3 Myc-E2F Regulatory Axis and DNA Metabolism and Replication

The regulation of specific subsets of Myc target genes is likely to be dependent on the context of Myc E-boxes in *cis*-regulatory modules that may contain five to eight different other transcription factor binding motifs. We have searched Myc genomic binding sequences for other DNA binding motifs with the hope of deciphering *cis*-regulatory modules of Myc target genes. Our in silico studies as well as recent global chromatin immunoprecipitation studies indicate that the E2F consensus binding sequence is significantly overrepresented among Myc-bound genomic sites, suggesting that a subset of Myc targets are likely to be coordinately regulated with E2F (Fig. 2). We found that nucleotide metabolic gene promoters are enriched with predicted Myc and E2F binding sites, indicating that these transcription factors could couple regulation of the cell cycle machinery with nucleotide metabolism. In

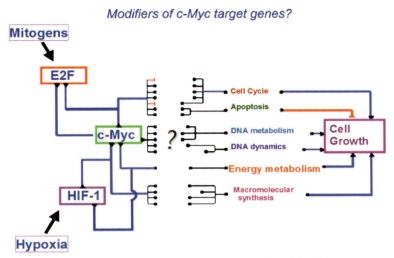

Fig. 2. The Myc target network affects approximately 15% of human genes. The cooperation between Myc and other transcription factors, such as E2F1 or HIF-1, is depicted to control subsets of Myc targets that are relevant under specific biological conditions, including mitogenic stimulation or hypoxic stress

fact, we found that virtually all genes encoding key enzymes in purine and pyrimidine metabolism are regulated directly by Myc and to a large extent bound by E2F1 (see next section). It is notable that Myc and E2F also have common target genes involved DNA replication (Archambault et al. 2005; Seo et al. 2005). The study of these transcriptional circuitries will reveal the complex network of transcription factors that connects cell metabolism to cell proliferation.

4   Effects of Antioxidants on Myc-Mediated Tumorigenesis, Reactive Oxygen Species and the Hypoxia-Inducible Factor

Both Myc and E2F1 have been implicated in the generation of reactive oxygen species (ROS) that could in turn contribute to tumorigenesis through triggering genomic instability (Tanaka et al. 2002; Vafa et al. 2002). Our preliminary findings of Myc and E2F1 regulation of target genes involved in DNA replication and nucleotide metabolism suggest the possibility that genomic instability induced by Myc and E2F1 could be linked to these target genes. In addition, Myc has been linked to the regulation of metabolic pathways occurring in the mitochondrion as well as mitochondrial biogenesis in *Drosophila* and mammalian systems (Li et al. 2005; Orian et al. 2003). Through both gain-of-function and loss-of-function studies, we found that Myc activity is associated with cellular mitochondrial mass and function (Li et al. 2005). Intriguingly, in a human B cell model with an inducible Myc construct, we found that Myc increases mitochondrial mass, cellular oxygen consumption and production of ROS. Recently, we demonstrated that a new Myc target, PGC1β, is directly linked to mitochondrial biogenesis (Zhang et al. 2007). This led us to question whether Myc-induced mitochondrial biogenesis could be linked to ROS production, in particular since ROS has been implicated in Myc-induced genomic instability (Dang et al. 2005).

We sought to determine the potential role of ROS in Myc-induced tumorigenesis and genomic instability and found that the antioxidant N-acetylcysteine (NAC) could dramatically diminish tumorigenesis of P493 human B cell xenografts in SCID mice as well as inhibiting a Myc-dependent transgenic model of hepatocellular carcinoma (Gao et al. 2007). We had expected that tumors arising in the P493 xenografts would display genomic instability, as seen in the transgenic murine lymphoma models (Karlsson et al. 2003), but surprisingly found through the use of spectral karyotyping and Illumina 300K SNP BeadChip genotyping that very few changes could be detected. These findings suggest that the antitumorigenic effects of NAC are not due to decreased genomic instability but rather to some other mechanism. Because of the dramatic phenotypic effects of NAC, we sought to determine whether tumor adaptation to the microenvironment could be affected, since the stability of the hypoxia-inducible factor HIF-1 is dependent on ROS metabolism (Kaelin 2005). Through these studies, we have identified HIF-1 as a major modifier of Myc function in vivo.

The activation of HIF, a transcription factor that is stabilized in response to hypoxia, significantly contributes to the induction of VEGF for angiogenesis and the conversion of glucose to lactate for tumor glucose metabolism (Brahimi-Horn et al. 2007). HIF-1 consists of an oxygen-sensitive HIF-1α subunit that heterodimerizes with HIF-1β to bind DNA. In high oxygen tension, HIF-1α is hydroxylated by prolyl hydroxylases (PHDs) using α-ketoglutarate derived from the Krebs cycle. The hydroxylated HIF-1α subunit is ubiquitylated by the von Hippel–Lindau protein, VHL, and destined for degradation by proteasomes, such that HIF-1α is continuously synthesized and degraded under nonhypoxic conditions (Semenza 2003). Hypoxia stabilizes HIF-1, and HIF-1 in turn directly transactivates glycolytic enzyme genes and VEGF. Hence, adaptation to the hypoxic tumor microenvironment results in increased glucose uptake and lactate production and angiogenesis.

There are several isoforms of PHDs, with PHD2 playing the most critical role in hydroxylating HIF-1α (Berra et al. 2006). The connection between mitochondrial ROS production, hypoxia and HIF-1 stabilization has emerged recently. In contrast to expectation, hypoxia or limited oxygen increases mitochondrial ROS rather diminishing it (Kaelin 2005). ROS, in turn, inactivates PHD2 and hence stabilizes HIF-1α. ROS inactivates PHDs through oxidation of the ferrous ion that is central and essential for the catalytic hydroxylation of prolines. Vitamin C has been shown to decrease HIF-1 levels through preventing the oxidation of the catalytic ferrous ion (Lu et al. 2005). Whether the antitumorigenic effects of vitamin C, which was claimed as a panacea for cancer therapy in the late 1970s and early 1980s, and that of NAC, which was touted to diminish genomic instability, are mediated through preventing genomic instability or adaptation to the tumor microenvironment has not been established (Cameron et al. 1979; Pauling et al. 1985). As seen from our studies, we have substantial evidence that antitumorigenic effects of NAC and vitamin C in Myc-mediated tumorigenesis is HIF-1 dependent, indicating that HIF is a critical in vivo Myc-modifier.

5   Collaboration of MYC and HIF in the Warburg Effect

Although HIF could attenuate the activity of endogenous MYC, when MYC is ectopically expressed, MYC cooperates with HIF (Fig. 2) to induce the expression of pyruvate dehydrogenase kinase 1 (PDK1), hexokinase 2 (HK2), and vascular endothelial growth factor (VEGFA) in human P493 B cells (Kim et al. 2007). Chromatin immunoprecipitation experiments indicate that these two factors do not inhibit one another from binding to the target genes. Hence, when MYC is overexpressed, it cooperates with HIF to stimulate glycolysis with increased production of lactate. This phenomenon is known as aerobic glycolysis, or the Warburg effect (Fig. 3). In contrast, anaerobic glycolysis, which is mediated largely by the transactivation of glycolytic genes by HIF-1, refers to the cellular consumption of glucose in hypoxia with an increased production of lactate from pyruvate. Under normal oxygen tension, pyruvate would normally be converted to acetyl-CoA by pyruvate dehydrogenase, whose function is inhibited by PDK1-mediated phosphorylation. Hence, the coordination between MYC and HIF in activating PDK1 contributes to the Warburg effect.

Over 80 years ago, Otto Warburg described that cancer tissues generally have increased glucose uptake with the propensity to convert glucose to lactate rather than to carbon dioxide through the use of oxidative phosphorylation (Warburg 1956). Warburg postulated that defective mitochondria contribute to the enhanced conversion of glucose to lactate in the presence of oxygen (Fig. 3). However, the evidence for defective mitochondria in cancers is only partly supported by mitochondrial DNA mutations that diminish oxidative phosphorylation, as specific subunits of the respiratory chain are encoded in the mitochondrial DNA sequence.

Because of the abnormal neovasculature, tumor cells exist in a hypoxic microenvironment and display increased glycolysis that could also be due to genetic alterations (Dewhirst et al. 2007; Gazit et al. 1997). The dramatic increase in glucose uptake by tumors provides a significant window between cancers and normal tissues for therapeutic targeting of glucose metabolism and for tumor imaging by positron emission tomography (PET) (Gatenby and Gillies 2004; Tatsumi et al. 2005). Aerobic glycolysis results from oncogenic genetic alterations that drive glycolysis and increase conversion of pyruvate to lactate. In fact, MYC, AKT and signal transduction pathways that stabilize HIF-1, such as loss of VHL or activation of RAS, can all contribute to the Warburg effect (Kim and Dang 2006; Plas and Thompson 2005) (Fig. 3).

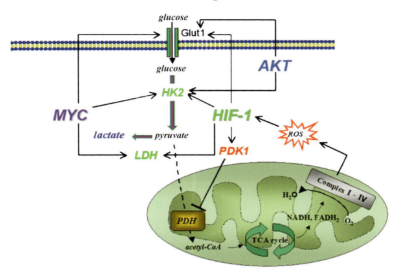

Fig. 3. The Warburg effect, or cancer aerobic glycolysis, which describes the propensity of cancers to take up glucose avidly and convert it to lactate, is depicted with molecular targets of MYC, HIF-1, and AKT. Glucose is transported by Glut1, which is activated by MYC, HIF, and AKT. Hexokinase 2 (*HK2*) is a transcriptional target of Myc and HIF and is a substrate of AKT; all three factors activate HK2, which phosphorylates and retains glucose intracellularly. Glucose is converted to pyruvate through glycolytic enzymes, many of which are common transcriptional target genes of MYC and HIF. Pyruvate is converted to lactate by lactate dehydrogenase (*LDH*), whose LDHA subunit is a common target of MYC and HIF. HIF (and MYC, not shown) transactivates pyruvate dehydrogenase 1 (*PDK1*), which phosphorylates and inactivates pyruvate dehydrogenase (*PDH*), which converts pyruvate to acetyl-CoA to drive the tricarboxylic acid (*TCA*) cycle. The TCA cycle feeds high-energy electrons to the transport chain—complexes I–IV—which produces ATP as well as reactive oxygen species (*ROS*), particularly under hypoxic conditions. ROS, in turn, contributes to the hypoxic stabilization of HIF-1

The cellular response to the hypoxic tumor microenvironment can also stabilize HIF-1, which stimulates glycolysis and lactate production.

The metabolic differences between cancer and normal tissues have been targeted for potential antitumor therapy. In particular, target genes that are common to MYC and HIF may be ideal, because they reflect both oncogenic activation and the adaptive response to tumor hypoxia. Lactate dehydrogenase A and pyruvate dehydrogenase kinase, which are both targets of HIF and MYC, could be attractive anticancer therapeutic targets (Bonnet et al. 2007; Fantin et al. 2006; Shim et al. 1997). In fact, small molecule inhibitors against LDH and PDK are already available, thus making the potential for translating these basic findings practicable. Hexokinase 2 is also another attractive therapeutic target because its expression is cooperatively increased by HIF and MYC, providing a wider therapeutic window between tumors and normal tissues (Mathupala et al. 2006). The transferrin receptor (TFRC) gene is also a target common to both MYC and HIF (O'Donnell et al. 2006; Tacchini et al. 1999), and small molecular inhibitors of TFRC have significant growth inhibitory activities (Kasibhatla et al. 2005; Pandey et al. 2007).

In summary, the oncogenic and hypoxic regulation of energy metabolism could contribute to the propensity of cancers to avidly take up glucose and convert it to lactate and provide a significant therapeutic window. For example, the induction of the glucose transporter, GLUT1, by both MYC and HIF provides a significant tumor-selective imaging window with PET scanning. It stands to reason that target genes that are common to MYC and HIF would be ideal therapeutic targets that distinguish between cancer and normal tissues. In this regard, a number of MYC and HIF targets are already being studied and exploited as potential anticancer therapeutic targets.

References

Adhikary S, Eilers M (2005) Transcriptional regulation and transformation by Myc proteins. Nat Rev Mol Cell Biol 6:635–645

Amati B, Alevizopoulos K, Vlach J (1998) Myc and the cell cycle. Front Biosci 3:D250–D268

Arabi A, Wu S, Ridderstrale K, Bierhoff H, Shiue C, Fatyol K, Fahlen S, Hydbring P, Soderberg O, Grummt I et al (2005) c-Myc associates with ribosomal DNA and activates RNA polymerase I transcription. Nat Cell Biol 7:303–310

Archambault V, Ikui AE, Drapkin BJ, Cross FR (2005) Disruption of mechanisms that prevent rereplication triggers a DNA damage response. Mol Cell Biol 25:6707–6721

Baudino TA, Cleveland JL (2001) The Max network gone mad. Mol Cell Biol 21:691–702

Bello-Fernandez C, Packham G, Cleveland JL (1993) The ornithine decarboxylase gene is a transcriptional target of c-Myc. Proc Natl Acad Sci U S A 90:7804–7808

Berra E, Ginouves A, Pouyssegur J (2006) The hypoxia-inducible-factor hydroxylases bring fresh air into hypoxia signalling. EMBO Rep 7:41–45

Blackwell TK, Kretzner L, Blackwood EM, Eisenman RN, Weintraub H (1990) Sequence-specific DNA binding by the c-Myc protein. Science 250:1149–1151

Blackwell TK, Huang J, Ma A, Kretzner L, Alt FW, Eisenman RN, Weintraub H (1993) Binding of myc proteins to canonical and noncanonical DNA sequences. Mol Cell Biol 13:5216–5224

Blackwood EM, Eisenman RN (1991) Max: a helix-loop-helix zipper protein that forms a sequence-specific DNA-binding complex with Myc. Science 251:1211–1217

Bonnet S, Archer SL, Allalunis-Turner J, Haromy A, Beaulieu C, Thompson R, Lee CT, Lopaschuk GD, Puttagunta L, Harry G et al (2007) A mitochondria-K+ channel axis is suppressed in cancer and its normalization promotes apoptosis and inhibits cancer growth. Cancer Cell 11:37–51

Boyd KE, Farnham PJ (1997) Myc versus USF: discrimination at the cad gene is determined by core promoter elements. Mol Cell Biol 17:2529–2537

Brahimi-Horn MC, Chiche J, Pouyssegur J (2007) Hypoxia signalling controls metabolic demand. Curr Opin Cell Biol 19:223–229

Burgin A, Bouchard C, Eilers M (1998) Control of cell proliferation by Myc proteins. Results Probl Cell Differ 22:181–197

Cameron E, Pauling L, Leibovitz B (1979) Ascorbic acid and cancer: a review. Cancer Res 39:663–681

Cole MD, McMahon SB (1999) The Myc oncoprotein: a critical evaluation of transactivation and target gene regulation. Oncogene 18:2916–2924

Cole MD, Nikiforov MA (2006) Transcriptional activation by the Myc oncoprotein. Curr Top Microbiol Immunol 302:33–50

Cory S, Harris AW, Langdon WY, Alexander WS, Corcoran LM, Palmiter RD, Pinkert CA, Brinster RL, Adams JM (1987) The myc oncogene and lymphoid neoplasia: from translocations to transgenic mice. Hamatol Bluttransfus 31:248–251

Cowling VH, Cole MD (2007) The Myc transactivation domain promotes global phosphorylation of the RNA polymerase II carboxy-terminal domain independently of direct DNA binding. Mol Cell Biol 27:2059–2073

Dang CV (1999) c-Myc target genes involved in cell growth, apoptosis and metabolism. Mol Cell Biol 19:1–11

Dang CV, McGuire M, Buckmire M, Lee WM (1989) Involvement of the 'leucine zipper' region in the oligomerization and transforming activity of human c-myc protein. Nature 337:664–666

Dang CV, Li F, Lee LA (2005) Could MYC induction of mitochondrial biogenesis be linked to ROS production and genomic instability? Cell Cycle 4:1465–1466

Dang CV, O'Donnell KA, Zeller KI, Nguyen T, Osthus RC, Li F (2006) The c-Myc target gene network. Semin Cancer Biol 16:253–264

Dewhirst MW, Cao Y, Li CY, Moeller B (2007) Exploring the role of HIF-1 in early angiogenesis and response to radiotherapy. Radiother Oncol 83:249–255

Eischen CM, Weber JD, Roussel MF, Sherr CJ, Cleveland JL (1999) Disruption of the ARF-Mdm2-p53 tumor suppressor pathway in Myc-induced lymphomagenesis. Genes Dev 13:2658–2669

Fantin VR, St-Pierre J, Leder P (2006) Attenuation of LDH-A expression uncovers a link between glycolysis, mitochondrial physiology and tumor maintenance. Cancer Cell 9:425–434

Felton-Edkins ZA, Kenneth NS, Brown TR, Daly NL, Gomez-Roman N, Grandori C, Eisenman RN, White RJ (2003) Direct regulation of RNA polymerase III transcription by RB, p53 and c-Myc. Cell Cycle 2:181–184

Fernandez PC, Frank SR, Wang L, Schroeder M, Liu S, Greene J, Cocito A, Amati B (2003) Genomic targets of the human c-Myc protein. Genes Dev 17:1115–1129

Gallant P, Shiio Y, Cheng PF, Parkhurst SM, Eisenman RN (1996) Myc and Max homologs in *Drosophila*. Science 274:1523–1527

Gao P, Zhang H, Dinavahi R, Li F, Xiang Y, Raman V, Bhujwalla ZM, Felsher DW, Cheng L, Pevsner J et al (2007) HIF-dependent antitumorigenic effect of antioxidants in vivo. Cancer Cell 12:230–238

Gartel AL, Ye X, Goufman E, Shianov P, Hay N, Najmabadi F, Tyner AL (2001) Myc represses the p21(WAF1/CIP1) promoter and interacts with Sp1/Sp3. Proc Natl Acad Sci U S A 98:4510–4515

Gatenby RA, Gillies RJ (2004) Why do cancers have high aerobic glycolysis? Nat Rev Cancer 4:891–899

Gazit Y, Baish JW, Safabakhsh N, Leunig M, Baxter LT, Jain RK (1997) Fractal characteristics of tumor vascular architecture during tumor growth and regression. Microcirculation 4:395–402

Gomez-Roman N, Grandori C, Eisenman RN, White RJ (2003) Direct activation of RNA polymerase III transcription by c-Myc. Nature 421:290–294

He L, Thomson JM, Hemann MT, Hernando-Monge E, Mu D, Goodson S, Powers S, Cordon-Cardo C, Lowe SW, Hannon GJ, Hammond SM (2005) A microRNA polycistron as a potential human oncogene. Nature 435:828–833

He TC, Sparks AB, Rago C, Hermeking H, Zawel L, da Costa LT, Morin PJ, Vogelstein B, Kinzler KW (1998) Identification of c-MYC as a target of the APC pathway. Science 281:1509–1512

Hemann MT, Bric A, Teruya-Feldstein J, Herbst A, Nilsson JA, Cordon-Cardo C, Cleveland JL, Tansey WP, Lowe SW (2005) Evasion of the p53 tumour surveillance network by tumour-derived MYC mutants. Nature 436:807–811

Hermeking H, Rago C, Schuhmacher M, Li Q, Barrett JF, Obaya AJ, O'Connell BC, Mateyak MK, Tam W, Kohlhuber F et al (2000) Identification of CDK4 as a target of c-MYC Proc Natl Acad Sci U S A 97:2229–2234

Hooker CW, Hurlin PJ (2006) Of myc and mnt. J Cell Sci 119:208–216

Iritani BM, Delrow J, Grandori C, Gomez I, Klacking M, Carlos LS, Eisenman RN (2002) Modulation of T-lymphocyte development, growth and cell size by the Myc antagonist and transcriptional repressor Mad1. EMBO J 21:4820–4830

Jamerson MH, Johnson MD, Dickson RB (2004) Of mice and Myc: c-Myc and mammary tumorigenesis. J Mammary Gland Biol Neoplasia 9:27–37

Johnston LA, Prober DA, Edgar BA, Eisenman RN, Gallant P (1999) Drosophila myc regulates cellular growth during development. Cell 98:779–790

Kaelin WG Jr (2005) ROS: really involved in oxygen sensing. Cell Metab 1:357–358

Karlsson A, Giuriato S, Tang F, Fung-Weier J, Levan G, Felsher DW (2003) Genomically complex lymphomas undergo sustained tumor regression upon MYC inactivation unless they acquire novel chromosomal translocations. Blood 101:2797–2803

Kasibhatla S, Jessen KA, Maliartchouk S, Wang JY, English NM, Drewe J, Qiu L, Archer SP, Ponce AE, Sirisoma N et al (2005) A role for transferrin receptor in triggering apoptosis when targeted with gambogic acid. Proc Natl Acad Sci U S A 102:12095–12100

Kato GJ, Barrett J, Villa-Garcia M, Dang CV (1990) An amino-terminal c-myc domain required for neoplastic transformation activates transcription. Mol Cell Biol 10:5914–5920

Kim JW, Dang CV (2006) Cancer's molecular sweet tooth and the Warburg effect. Cancer Res 66:8927–8930

Kim JW, Gao P, Liu YC, Semenza GL, Dang CV (2007) Hypoxia-inducible factor 1 and dysregulated c-Myc cooperatively induce vascular endothelial growth factor and metabolic switches hexokinase 2 and pyruvate dehydrogenase kinase 1. Mol Cell Biol 27:7381–7393

Kim S, Li Q, Dang CV, Lee LA (2000) Induction of ribosomal genes and hepatocyte hypertrophy by adenovirus-mediated expression of c-Myc in vivo. Proc Natl Acad Sci U S A 97:11198–11202

Koshiji M, To KK, Hammer S, Kumamoto K, Harris AL, Modrich P, Huang LE (2005) HIF-1alpha induces genetic instability by transcriptionally downregulating MutSalpha expression. Mol Cell 17:793–803

Langdon WY, Harris AW, Cory S, Adams JM (1986) The c-myc oncogene perturbs B lymphocyte development in E-mu-myc transgenic mice. Cell 47:11–18

Leder A, Pattengale PK, Kuo A, Stewart TA, Leder P (1986) Consequences of widespread deregulation of the c-myc gene in transgenic mice: multiple neoplasms and normal development. Cell 45:485–495

Leone G, Sears R, Huang E, Rempel R, Nuckolls F, Park CH, Giangrande P, Wu L, Saavedra HI, Field SJ et al (2001) Myc requires distinct E2F activities to induce S phase and apoptosis. Mol Cell 8:105–113

Li F, Wang Y, Zeller KI, Potter JJ, Wonsey DR, O'Donnell KA, Kim JW, Yustein JT, Lee LA, Dang CV (2005) Myc stimulates nuclearly encoded mitochondrial genes and mitochondrial biogenesis. Mol Cell Biol 25:6225–6234

Liao DJ, Dickson RB (2000) c-Myc in breast cancer. Endocr Relat Cancer 7:143–164

Lu H, Dalgard CL, Mohyeldin A, McFate T, Tait AS, Verma A (2005) Reversible inactivation of HIF-1 prolyl hydroxylases allows cell metabolism to control basal HIF-1. J Biol Chem 280:41928–41939

Luo Z, Saha AK, Xiang X, Ruderman NB (2005) AMPK, the metabolic syndrome and cancer. Trends Pharmacol Sci 26:69–76

Marhin WW, Chen S, Facchini LM, Fornace AJ Jr, Penn LZ (1997) Myc represses the growth arrest gene gadd45. Oncogene 14:2825–2834

Mateyak MK, Obaya AJ, Sedivy JM (1999) c-Myc regulates cyclin D-Cdk4 and -Cdk6 activity but affects cell cycle progression at multiple independent points. Mol Cell Biol 19:4672–4683

The Interplay Between MYC and HIF in the Warburg Effect 51

Mathupala SP, Ko YH, Pedersen PL (2006) Hexokinase II: cancer's double-edged sword acting as both facilitator and gatekeeper of malignancy when bound to mitochondria. Oncogene 25:4777–4786

McArthur GA, Laherty CD, Queva C, Hurlin PJ, Loo L, James L, Grandori C, Gallant P, Shiio Y, Hokanson WC et al (1998) The Mad protein family links transcriptional repression to cell differentiation. Cold Spring Harb Symp Quant Biol 63:423–433

Miliani de Marval PL, Macias E, Rounbehler R, Sicinski P, Kiyokawa H, Johnson DG, Conti CJ, Rodriguez-Puebla ML (2004) Lack of cyclin-dependent kinase 4 inhibits c-myc tumorigenic activities in epithelial tissues. Mol Cell Biol 24:7538–7547

Nikiforov MA, Chandriani S, O'Connell B, Petrenko O, Kotenko I, Beavis A, Sedivy JM, Cole MD (2002) A functional screen for Myc-responsive genes reveals serine hydroxymethyltransferase, a major source of the one-carbon unit for cell metabolism. Mol Cell Biol 22:5793–5800

O'Donnell KA, Wentzel EA, Zeller KI, Dang CV, Mendell JT (2005) c-Myc-regulated microRNAs modulate E2F1 expression. Nature 435:839–843

O'Donnell KA, Yu D, Zeller KI, Kim JW, Racke F, Thomas-Tikhonenko A, Dang CV (2006) Activation of transferrin receptor 1 by c-Myc enhances cellular proliferation and tumorigenesis. Mol Cell Biol 26:2373–2386

Orian A, van Steensel B, Delrow J, Bussemaker HJ, Li L, Sawado T, Williams E, Loo LW, Cowley SM, Yost C et al (2003) Genomic binding by the *Drosophila* Myc Max Mad/Mnt transcription factor network. Genes Dev 17:1101–1114

Oskarsson T, Trumpp A (2005) The Myc trilogy: lord of RNA polymerases. Nat Cell Biol 7:215–217

Pandey MK, Sung B, Ahn KS, Kunnumakkara AB, Chaturvedi MM, Aggarwal BB (2007) Gambogic acid, a novel ligand for transferrin receptor, potentiates TNF-induced apoptosis through modulation of the nuclear factor-{kappa}B signaling pathway. Blood 110:3517–3525

Pauling L, Nixon JC, Stitt F, Marcuson R, Dunham WB, Barth R, Bensch K, Herman ZS, Blaisdell BE, Tsao C et al (1985) Effect of dietary ascorbic acid on the incidence of spontaneous mammary tumors in RIII mice. Proc Natl Acad Sci U S A 82:5185–5189

Pelengaris S, Littlewood T, Khan M, Elia G, Evan G (1999) Reversible activation of c-Myc in skin: induction of a complex neoplastic phenotype by a single oncogenic lesion. Mol Cell 3:565–577

Plas DR, Thompson CB (2005) Akt-dependent transformation: there is more to growth than just surviving. Oncogene 24:7435–7442

Prendergast GC, Ziff EB (1991) Methylation-sensitive sequence-specific DNA binding by the c-Myc basic region. Science 251:186–189

Prendergast GC, Lawe D, Ziff EB (1991) Association of Myn, the murine homolog of max, with c-Myc stimulates methylation-sensitive DNA binding and ras cotransformation. Cell 65:395–407

Schneider A, Peukert K, Eilers M, Hanel F (1997) Association of Myc with the zinc-finger protein Miz-1 defines a novel pathway for gene regulation by Myc. Curr Top Microbiol Immunol 224:137–146

Schreiber-Agus N, Stein D, Chen K, Goltz JS, Stevens L, DePinho RA (1997) Drosophila Myc is oncogenic in mammalian cells and plays a role in the diminutive phenotype. Proc Natl Acad Sci U S A 94:1235–1240

Sears RC, Nevins JR (2002) Signaling networks that link cell proliferation and cell fate. J Biol Chem 277:11617–11620

Semenza GL (2003) Targeting HIF-1 for cancer therapy. Nat Rev Cancer 3:721–732

Seo J, Chung YS, Sharma GG, Moon E, Burack WR, Pandita TK, Choi K (2005) Cdt1 transgenic mice develop lymphoblastic lymphoma in the absence of p53. Oncogene 24:8176–8186

Shchors K, Shchors E, Rostker F, Lawlor ER, Brown-Swigart L, Evan GI (2006) The Myc-dependent angiogenic switch in tumors is mediated by interleukin 1beta. Genes Dev 20:2527–2538

Shim H, Dolde C, Lewis BC, Wu CS, Dang G, Jungmann RA, Dalla-Favera R, Dang CV (1997) c-Myc transactivation of LDH-A: implications for tumor metabolism and growth. Proc Natl Acad Sci U S A 94:6658–6663

Tacchini L, Bianchi L, Bernelli-Zazzera A, Cairo G (1999) Transferrin receptor induction by hypoxia. HIF-1-mediated transcriptional activation and cell-specific post-transcriptional regulation. J Biol Chem 274:24142–24146

Tanaka H, Matsumura I, Ezoe S, Satoh Y, Sakamaki T, Albanese C, Machii T, Pestell RG, Kanakura Y (2002) E2F1 and c-Myc potentiate apoptosis through inhibition of NF-kappaB activity that facilitates MnSOD-mediated ROS elimination. Mol Cell 9:1017–1029

Tatsumi M, Cohade C, Nakamoto Y, Fishman EK, Wahl RL (2005) Direct comparison of FDG PET, CT findings in patients with lymphoma: initial experience. Radiology 237:1038–1045

Trumpp A, Refaeli Y, Oskarsson T, Gasser S, Murphy M, Martin GR, Bishop JM (2001) c-Myc regulates mammalian body size by controlling cell number but not cell size. Nature 414:768–773

Vafa O, Wade M, Kern S, Beeche M, Pandita TK, Hampton GM, Wahl GM (2002) c-Myc can induce DNA damage, increase reactive oxygen species and mitigate p53 function: a mechanism for oncogene-induced genetic instability. Mol Cell 9:1031–1044

Warburg O (1956) On the origin of cancer cells. Science 123:309–314

Wu S, Cetinkaya C, Munoz-Alonso MJ, von der Lehr N, Bahram F, Beuger V, Eilers M, Leon J, Larsson LG (2003) Myc represses differentiation-induced p21CIP1 expression via Miz-1-dependent interaction with the p21 core promoter. Oncogene 22:351–360

Zhang H, Gao P, Fukuda R, Kumar G, Krishnamachary B, Zeller KI, Dang CV, Semenza GL (2007) HIF-1 inhibits mitochondrial biogenesis and cellular respiration in VHL-deficient renal cell carcinoma by repression of C-MYC activity. Cancer Cell 11:407–420

Zindy F, Eischen CM, Randle DH, Kamijo T, Cleveland JL, Sherr CJ, Roussel MF (1998) Myc signaling via the ARF tumor suppressor regulates p53-dependent apoptosis and immortalization. Genes Dev 12:2424–2433

Ernst Schering Foundation Symposium Proceedings, Vol. 4, pp. 55–78
DOI 10.1007/2789_2008_089
© Springer-Verlag Berlin Heidelberg
Published Online: 03 July 2008

Using Metabolomics to Monitor Anticancer Drugs

Y.-L. Chung, J.R. Griffiths[✉]

Cancer Research UK, Li Ka Shing Centre, Cancer Research UK Cambridge Research
Institute, Robinson Way, CB2 0RE Cambridge, UK
email: *John.Griffiths@cancer.org.uk*

1	Introduction	56
2	Pharmacodynamic Markers	58
3	Conventional Cytotoxic Drugs	59
3.1	5-Fluorouracil	59
3.2	Ifosfamide	59
3.3	Cyclophosphamide	61
4	Cyclin-Kinase Inhibitor	61
5	HSP90 Inhibitor	61
6	Choline Kinase Inhibitor	62
7	HDAC Inhibitors	63
7.1	LAQ824	63
7.2	SAHA	64
7.3	Phenylbutyrate	64
8	Vascular Disruption Agents	64
8.1	DMXAA	65
8.2	ZD6126	65
8.3	Combretastatin A4 Phosphate	65
9	HIF-1α Inhibitor	66
10	PI3K Inhibitor	66
10.1	LY294002 and Wortmannin	66
10.2	PI103	67
11	MAPK Inhibitor	67
12	Fatty Acid Synthase Inhibitor	67
13	Antimicrotubule Drug	68

| 14 | Discussion | 68 |
| References | | 75 |

Abstract. The metabolome of a cancer cell is likely to show changes after responding to an anticancer drug. These changes could be used to decide whether to continue treatment or, in the context of a drug trial, to indicate whether the drug is working and perhaps its mechanism of action. (Nuclear) magnetic resonance spectroscopy (NMR/MRS) methods can offer important insights into novel anticancer agents in order to accelerate the drug development process including time-course studies on the effect of a drug on its site of action (termed pharmacodynamics), in this case the cancer. In addition, some classes of anticancer agents currently under development (e.g. antiangiogenics) are designed to be used in combination with other drugs and will not cause tumour shrinkage when used as single agents in Phase 1 clinical trials. Thus NMR/MRS may have a special role in monitoring the pharmacodynamic actions of such drugs in early-phase clinical trials. This review focuses on the use of ex vivo NMR and in vivo MRS methods for monitoring the effect of some novel anticancer drugs on the cancer metabolome. Ex vivo NMR methods are complementary to in vivo measurements, as they can provide additional information and help in the interpretation of the in vivo data.

1 Introduction

Metabolomics is the study of the totality of small-molecule metabolites in an organism, cell or disease state. Unlike the genome, the metabolome of a cell can change from minute to minute, depending on factors such as its stage in the cell cycle or its environment. Similarly, when a cancer cell responds to an anticancer drug its metabolome is likely to show changes that could be used to decide whether to continue treatment or, in the context of a drug trial, to indicate whether the drug is working and perhaps its mechanism of action. In this review, we will focus on the use of (nuclear) magnetic resonance spectroscopy (NMR/MRS) methods for monitoring the effect of anticancer drugs on the cancer metabolome. Following the standard convention, the nomenclature we will use is as follows. The term "NMR" will be used for the magnetic resonance spectroscopy technique when used ex vivo on tissue extracts, biopsies, etc. The term "MRS" will be used to denote studies by the same technique on living animals or patients.

Conventional high-resolution nuclear magnetic resonance (NMR) can be used ex vivo to analyse cultured cells or biopsies from tumours, either by first extracting the sample into perchloric acid or chloroform/methanol or by high-resolution magic angle spinning (HR-MAS) NMR of solid samples. It is also possible to use magnetic resonance spectroscopy (MRS) noninvasively in vivo to obtain spectra from tumours in living animals or patients (note that when NMR-based methods are used in vivo it is conventional to drop the word "nuclear"). This ability to measure metabolites repeatedly and noninvasively in a living subject is a unique advantage of MRS, and although the number of metabolites detected is small, this method can be exploited in metabolomics.

In fact, all NMR and MRS methods are inherently insensitive and detect only a relatively small fraction of the metabolome. However, they have several compensating advantages in comparison with the more usual mass spectrometry or gas chromatography methods. When conventional high-resolution NMR is used on cultured cells or biopsies, the fact that it is not necessary to derivatize the metabolites within the sample, or to ionize them, removes two major sources of imprecision that impair the quantitative use of mass spectrometric data. Consequently the relative concentrations of the metabolites in a sample can be established by ex vivo NMR with a high degree of precision, making it easy to detect differences between pre- and post-drug spectra. When MRS is used in vivo on tumours in patients or experimental animals, it has an additional advantage, since repeated spectra can be obtained, giving a time-course. Time-course studies on the effect of a drug on its site of action – in this case the cancer – are termed pharmacodynamics.

Most MRS work on pharmacodynamics in vivo has been done using ^{1}H and/or ^{31}P MRS, while ex vivo studies with high-resolution NMR on cell or tissue extracts or by HR-MAS NMR on solid tissue biopsies are usually conducted by ^{1}H NMR, although ^{31}P NMR is sometimes also used. In vivo ^{31}P MRS can be used to obtain markers for tissue bioenergetics (nucleotide triphosphate [NTP], inorganic phosphate [Pi]), intracellular pH (pHi) and membrane turnover (phosphorus-containing components of phospholipid membrane metabolism: phosphomonoester [PME] and phosphodiester [PDE] compounds). The composition of the PME and PDE peaks was established by in vitro ^{31}P NMR of extracts of cancer cells or tumours, since these metabolite peaks are more readily resolved in vitro than in vivo (Evanochko et al. 1984; de Certaines et al. 1993). In tumours, PMEs are made up mainly of phosphocholine (PC) and phosphoethanolamine (PE), which are precursors of the membrane phospholipids phosphatidylcholine (PtdC) and phosphatidylethanolamine (PtdE). The PDEs are comprised of glycerophosphocholine (GPC) and glycerophosphoethanolamine (GPE), breakdown products of PtdC and PtdE, respectively. Thus, although membrane components such as PtdC and PtdE give peaks that are too broad to be detected, one can monitor their precursors and breakdown products. In addition, by ^{1}H MRS in vivo it is possible to detect metabolites such as lactate, lipids and a peak that is usually assigned to total choline-containing compounds (tCho) (Shungu et al. 1992). The choline signals in the tCho peak are mainly from free choline, PC and GPC, but resonances from myo-inositol and taurine (Sitter et al. 2002); and from PE (Govindaraju et al. 2000) and GPE (Nelson et al. 1996) are also present in this region.

2   Pharmacodynamic Markers

Most of the published pharmacodynamic work in cancer has been done on routinely used chemotherapeutic drugs, but MRS can also offer important insights into novel anticancer agents in order to accelerate the drug development process. In addition, some classes of anticancer agents currently under development (e.g. antiangiogenics) are designed to be used in combination with other drugs and may not cause tumour shrinkage when used as single agents in Phase 1 clinical trials. Thus, MRS may have a special role in monitoring the pharmacodynamic actions of such drugs in early-phase clinical trials.

The following section demonstrates the use of metabolomic methods based on in vivo MRS and in vitro ^{1}H and ^{31}P NMR to assess tumour response to selected examples of conventional cytotoxic agents and to some novel drugs with specific molecular targets. MR-based biomarkers such as these could potentially provide surrogate pharmacodynamic markers for use in clinic trials.

Figure 1 shows some different patterns of metabolic change that are found by ^{31}P MRS in tumours following treatment. The examples are drawn from four different classes of novel anticancer drug. Each of these drugs induces a different pattern of change in the ^{31}P MR spectrum (indicated by the asterisks over the peaks), reflecting different alterations in the subset of the metabolome that is detected by this analytical modality. Although single spectra are shown in Fig. 1, the alterations in the peak areas indicated by the asterisks were found to be statistically significant in larger experiments.

3 Conventional Cytotoxic Drugs

3.1 5-Fluorouracil

5-Fluorouracil (5-FU) is an antimetabolite drug that is widely used in medical oncology (Chen and Grem 1992). In vivo and in vitro ^{31}P MRS were used to examine the pharmacodynamic effect of 5-FU on a mouse mammary carcinoma model (Street et al. 1997). Increased NTP to Pi and PCr to Pi ratios were observed in vivo 48 h after 5-FU treatment, implying an improvement in tumour energy metabolism. This increase in high-energy phosphate metabolites relative to Pi (which is also seen in response to a number of other anticancer drugs; see Sects. 3.2 and 3.3) could be due to a decrease in cell number allowing more effective access of oxygen and dissolved nutrients to the remaining cells. The PE to PC ratio was also elevated after 5FU treatment; this was found to be due to an increase in PE (as confirmed by in vitro ^{31}P NMR). In addition, increases in GPC and GPE were also observed (Street et al. 1997).

3.2 Ifosfamide

Tumour growth inhibition was observed in a mouse xenograft model of paediatric embryonal rhabdomyosarcoma (Rd), 7 days after ifosfamide treatment. ^{31}P MRS of Rd tumours in vivo showed significant increases in PME to Pi and β-ATP to Pi ratios after ifosfamide treatment when compared with pretreatment values. The rise in the PME to Pi ratio is

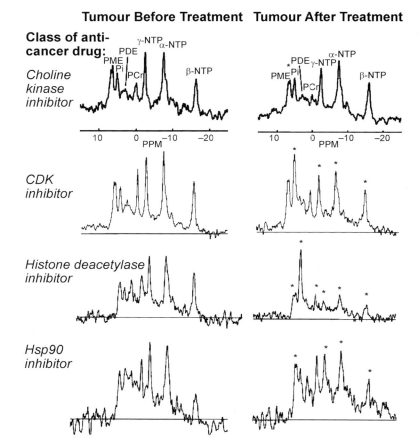

*showing metabolites that have changed after treatment.

Fig. 1. In vivo ^{31}P MRS profile of HT29 tumours following treatment with examples of four different classes of novel anticancer drug. The examples are: MN58b—choline kinase inhibitor (Al-Saffar et al. 2006); CYC202—CDK inhibitor (Troy et al. 2002); LAQ824—histone deacetylase inhibitor (Chung et al. 2007); 17-AAG—Hsp90 inhibitor (Chung et al. 2003)

due to an increase in PE and unchanged Pi levels (as confirmed by in vitro ^{31}P NMR of tumour extracts) (Vaidya et al. 2003).

3.3 Cyclophosphamide

Increases in the NTP to Pi and PCr to Pi ratios were found in vivo in cyclophosphamide-treated mouse mammary carcinomas. The PE to PC ratio was also elevated after treatment, and this effect was found to be due to a drop in PC (as confirmed by in vitro ^{31}P NMR of tumour extracts). In addition, rises in GPC and GPE levels were also found in extracts of cyclophosphamide-treated tumours when compared with controls (Street et al. 1995).

4 Cyclin-Kinase Inhibitor

CYC202 (R-roscovitine) inhibits the cyclin-dependent kinases 1, 2 and 7, and thus blocks cell cycle progression. Its action in vivo was monitored by ^{31}P MRS in human colon xenografts to gain insights into the biochemical changes associated with cell cycle disruption. Following 4 days of CYC202 treatment, in vivo ^{31}P MRS of HT29 tumours showed a significant decrease in intracellular pH, and the NTP to total phosphorus signal (TotP) ratio, and increases in the Pi to TotP and Pi to NTP ratios. In vitro ^{1}H NMR of CYC202-treated tumour extracts showed falls in GPC, glycine and glutamate when compared with controls. CYC202 treatment caused reduction in tumour proliferation and tissue pH, and impairment in tumour bioenergetics (Chung et al. 2002).

5 HSP90 Inhibitor

The heat-shock protein HSP90 is of interest as an anticancer target (Necker 2002) because it helps maintain the shape of many oncogenic proteins. Inhibiting single oncogenes with "magic bullet" drugs has proved somewhat disappointing as the cancers often become resistant, so a "magic shotgun" that hits many oncogenes simultaneously is an attractive concept. The HSP90 inhibitor 17AAG is currently in clinical trial.

The actions of 17AAG were monitored on several cultured human colon cancer cell lines (HCT116, HT29 and SW620) and on a human colon tumour xenograft model (HT29) (Chung et al. 2003). In the HCT116, HT29 and SW620 cell lines, there were significant increases in PC and GPC. In vivo, after 4 days of 17AAG treatment, the HT29 tumours showed significant growth delay as well as increased ratios of PME to PDE, PME to TotP and PME to β-NTP, and a decrease in the β-NTP to TotP ratio. The rises in the PME ratios were due to increases in PC and PE, as confirmed by in vitro ^{1}H and ^{31}P NMR studies of 17AAG-treated tumour extracts when compared with controls. A significant inverse correlation was found between the percentage change in the PME to PDE ratio and the percentage change in tumour size following 17AAG treatment (Chung et al. 2003)

6   Choline Kinase Inhibitor

Choline kinase (ChoK) is a cytosolic enzyme that catalyses the phosphorylation of choline to form PC, which is involved in cell membrane synthesis. Elevated levels of PC and ChoK found in tumours are associated with cell proliferation and malignant transformation. MN58b is an inhibitor of ChoK.

A significant growth delay was observed in the MN58b-treated HT29 xenografts when compared with controls. In vivo, ^{31}P and ^{1}H MRS of the HT29 xenografts showed a decrease in the ratio of PME to TotP and a decrease in total choline (tCho) concentration after 5 days of MN58b treatment. Extracts of drug-treated HT29 tumours showed significant decreases in PC when compared with controls. No changes in the other phospholipid metabolites (PE, GPC and GPE) were observed. Similar metabolite changes to those in HT29 tumours were also found in MN58b-treated MDA-MB-231 tumours (Al-Saffar et al. 2006). A drop in PC was also found in MN58b-treated HT29 cell extracts when compared with controls (Al-Saffar et al. 2006).

7 HDAC Inhibitors

7.1 LAQ824

LAQ824 is a novel anticancer drug that inhibits histone deacetylase (HDAC), resulting in growth inhibition, cell cycle arrest and apoptosis. Significant tumour growth inhibition was observed in HT29 xenografts following 2 days of LAQ824 treatment when compared with vehicle-treated controls. In vivo, the ratio of PME to TotP was significantly increased in LAQ824-treated HT29 xenografts and this ratio was found to correlate inversely with tumour response. This PME increase is confirmed by the significant rises in PC and PE levels observed in ^{1}H- and ^{31}P-NMR spectra of LAQ824-treated tumour extracts when compared with controls. These increases in PC and PME metabolites could potentially be used as biomarkers of HDAC inhibition (Chung et al. 2007).

Marked decreases in NTP and PCr, and an increase in Pi were also found in vivo by ^{31}P-MRS of LAQ824-treated tumours; in addition, significant decreases in intracellular pH, and in the β-NTP to TotP and β-NTP to Pi ratios, and an increase in Pi to TotP ratio were observed. These observations indicate that the tumour's bioenergetics are severely compromised following treatment, in contrast to the rises in bioenergetic state observed with 5-FU, ifosfamide and cyclophosphamide (see Sects. 3.1, 3.2, and 3.3). Elevated free choline, leucine, iso-leucine and valine levels and reduced GPC, GPE, glutamate, glutamine, glucose PCr and creatine levels were found in LAQ824-treated HT29 tumour extracts when compared with controls. These metabolite changes are also consistent with impaired tumour bioenergetics. A marked reduction of CD31 staining was found in LAQ824-treated tumours, indicating reduced vessel density in the LAQ824-treated group when compared with controls. The metabolite changes found in treated tumour extracts, the vascular changes and the effects that were observed in tumour bioenergetics, are consistent with the known antiangiogenic effect (Qian et al. 2004, and see Sect. 8.1) of LAQ824 on solid tumours (Chung et al. 2007).

Inhibition of HDAC by LAQ824 resulted in altered phospholipid metabolism and compromised tumour bioenergetics. The changes in phospholipid metabolism might function as noninvasive biomarkers of

HDAC inhibition per se, whereas the alterations in energy-associated metabolites could be used as biomarkers of the drug's antiangiogenic effects (Chung et al. 2007).

7.2 SAHA

SAHA is an anticancer drug that acts by inhibition of histone deacetylase (HDAC). Sankaranarayanapillai et al. used in vivo ^{1}H MRS and in vitro ^{31}P and ^{13}C NMR to study the effects of SAHA on a prostate cancer line (PC3). Increased PC and total choline levels were found in SAHA-treated cell extracts and these changes were inversely correlated with HDAC activity (Sankaranarayanapillai et al. 2005, 2007a). Increases in tCho to total signal and tCho to lipid ratios were found in PC3 tumours after 2 days of SAHA treatment. However, tCho normalized to the internal water signal remained unchanged, in contrast to the cell extract data (Sankaranarayanapillai et al. 2007b).

7.3 Phenylbutyrate

Phenylbutyrate is also an HDAC inhibitor. Increases in GPC, tCho and NMR-visible lipids were found in DU145 human prostatic carcinoma cells treated with phenylbutyrate. These effects were accompanied by significant increases in cytoplasmic lipid droplets and intracellular lipid volume fraction as observed by morphometric analysis of Oil Red O-stained cells. Phenylbutyrate treatment of cells caused cell cycle arrest in the G1 phase and induction of apoptosis. The simultaneous accumulation of mobile lipid and GPC suggests that phenylbutyrate induces phospholipid catabolism via a phospholipase-mediated pathway (Milkevitch et al. 2005).

8 Vascular Disruption Agents

Vascular disruption agents (VDAs) are novel anticancer drugs that destroy the blood vessels supplying a tumour, eventually causing massive necrosis.

8.1 DMXAA

DMXAA is a VDA. Following treatment with DMXAA, HT29 tumours showed dose-dependent decreases in both β-NTP to Pi and PDE to PME 6 h after treatment, when compared with vehicle-controls (MacPhail et al. 2005). A significant decrease in tCho in vivo was found 24 h after treatment with 21 mg/kg DMXAA; this was associated with a significant reduction in the concentration of the membrane degradation products GPE and GPC measured in tissue extracts. Elevated free choline was found in DMXAA-treated tumour extracts when compared with vehicle controls. These reductions in tumour energetics and membrane turnover are consistent with the vascular-disrupting activity of DMXAA. In vivo ^{31}P MRS revealed tumour response to DMXAA at doses below the maximum tolerated dose for mice (MacPhail et al. 2005), so this method might have use as a surrogate biomarker for this class of agent.

8.2 ZD6126

Radiation-induced fibrosarcoma 1 (RIF-1) tumours treated with ZD6126, another VDA, showed a significant reduction in tCho in vivo, 24 h after treatment, whereas vehicle-treated control tumours showed a significant increase in tCho. Ex vivo HRMAS NMR of tumour tissues and ^{1}H NMR of tumour extracts revealed significant reductions in PC and GPC in ZD6126-treated tumours; this confirmed the in vivo tCho finding. ZD6126-induced reduction of the amount of choline-containing compounds is consistent with a reduction in cell membrane turnover associated with necrosis and cell death following disruption of the tumour vasculature (Madhu et al. 2006).

8.3 Combretastatin A4 Phosphate

Combretastatin A4 phosphate is a VDA. Significant drops in the β-NTP to Pi ratio and intracellular pH were observed in C3H murine mammary tumours 1 h after combretastatin A4 phosphate treatment. An increase in lactate level was also found after treatment, but this effect was not observed consistently. The reduction in tumour energetics and pH was consistent with a reduction in tumour blood flow, but this occurred before any significant incidence of haemorrhagic necrosis was detected (Maxwell et al. 1998).

The acute effects of the antivascular drug combretastatin A4 phosphate were further investigated by Beauregard et al. The tumour bioenergetics of five tumour models – LoVo, RIF-1, SaS, SaF and HT29 – were examined by in vivo ^{31}P MRS following treatment with combretastatin A4 phosphate. A significant increase in the Pi to NTP ratio was observed by in vivo localized ^{31}P MRS in LoVo and RIF-1 tumours 3 h after treatment. SaS, SaF and HT29 tumours did not respond to the same degree. This tumour susceptibility to combretastatin A4 phosphate was found to correlate with vascular permeability (Beauregard et al. 2001).

9 HIF-1α Inhibitor

The hypoxia-inducible transcription factor HIF-1 plays an important role in the development of many tumours. PX-478 is an inhibitor of HIF-1α, one of the two subunits of the HIF-1 protein. In vivo ^{1}H MRS showed a significant decrease of the tCho in HT29 xenografts after 12 and 24 h of PX478 treatment. These changes were due to decreases in PC and GPC, as confirmed by high-resolution ^{1}H and ^{31}P NMR of tumour extracts. Reductions of PE, GPE and myo-inositol were also found in PX-478-treated tumour extracts when compared with controls. Significant reductions in cardiolipin, PtdE and phosphatidylinositol (PtdI) were also observed in lipid extracts of tumours after PX478 treatment when compared with vehicle controls. The significant change in tCho could potentially be used as an in vivo MRS biomarker for drug response following HIF-1α inhibition. The in vitro metabolic profiles of tumour indicated that GPC, PC, myo-inositol, PE, GPE, CL, PtdE and PtdI are potential ex vivo response biomarkers (Jordan et al. 2005).

10 PI3K Inhibitor

10.1 LY294002 and Wortmannin

LY294002 and Wortmannin are inhibitors of the PI3 kinase (PI3K) pathway. In vitro ^{31}P NMR of MDA-MB-231, MCF-7, and Hs578T cell

Using Metabolomics to Monitor Anticancer Drugs

extracts showed significant decreases in PC and increases in GPC levels following LY294002 treatment. A significant drop in the NTP level was also found in Hs578 cells following treatment, but not in MCF-7 or MDA-MB-231 cells. The drop in PC and rise in GPC levels were also observed by ^{31}P NMR of intact MDA-MB-231 cells following exposure to LY294002. A significant decrease in PC was also observed in extracts of MDA-MB-231 cell following treatment with another PI3K inhibitor, wortmannin, and no significant changes in the other metabolite levels were found. This study indicates that PI3K inhibition in human breast cancer cells by LY294002 and wortmannin is associated with a decrease in PC levels (Beloueche-Babari et al. 2006).

10.2 PI103

Treatment of PC3 cells with PI103, a PI3K inhibitor, caused a dose- and time-dependent decrease in PC, PE and NTP levels. These metabolite changes were associated with the drop in AKT phosphorylation and choline kinase activity (Al-Saffar et al. 2007).

11 MAPK Inhibitor

U0126 is a mitogen-activated protein kinase (MAPK) signalling inhibitor. Treatment of MDA-MB-231, MCF-7 and Hs578 cells with U0126 caused inhibition of extracellular signal-regulated kinase (ERK1/2) phosphorylation and a significant drop in PC, as shown by ^{31}P NMR of cell extracts. Similar changes were also observed in colon carcinoma HCT116 cells following exposure to U0126. The reductions in PC level in MDA-MB-231 and HCT116 cells were significantly correlated with the drop in P-ERK1/2 levels. This study showed that MAPK signalling inhibition with U0126 is associated with a time-dependent decrease in cellular PC levels (Beloueche-Babari et al. 2005).

12 Fatty Acid Synthase Inhibitor

Fatty acid synthase (FASE) is a key enzyme that catalyses the terminal steps in the synthesis of saturated fatty acids. FASE expression is low in normal human tissues because most lipids are obtained from the diet. Over-expression of FASE has been found in a wide variety of human cancers and is associated with a poor prognosis. Hence, FASE is an attractive therapeutic target for developing novel anticancer drugs such as orlistat, a fatty acid synthase inhibitor.

Treatment of PC3 cells with orlistat caused a drop in FASE activity and inhibition of cell proliferation (Ross et al. 2007). In vitro ^{31}P and ^{1}H NMR of cell extracts and ^{13}C NMR of extracts of cells treated with labelled choline show reduced levels of fatty acids, PtdC and PC following orlistat treatment. These data indicated the inhibition of de novo synthesis of these metabolites after treatment. Correlations were found between inhibition of FASE and inhibition of de novo synthesis of fatty acids, PtdC and PC (Ross et al. 2007).

13 Antimicrotubule Drug

Docetaxel is an antimicrotubule agent. Significant decreases in tumour PC levels were observed in two breast tumour models, MCF-7 and MDA-MB-231, 2–4 days after docetaxel treatment. A significant decrease in PC was found in vivo after docetaxel treatment, and this observation was confirmed by in vitro NMR of tumour extracts. An increase in GPC was also found in docetaxel-treated tumour extracts. These changes occurred in parallel with tumour growth delay, cell-cycle arrest and cell death (Morse et al. 2007). Since PC is a precursor and GPC is a breakdown product of PtdC in phospholipid membranes, these results would be consistent with a decreased synthesis and increased degradation of cell membranes.

14 Discussion

Table 1 summarizes some of the responses of cancer cells and tumours to the drugs mentioned in this review. The responses highlighted here are those that might be used in vivo, and might therefore be useful for monitoring clinical trials. They comprise cellular bioenergetic parameters that can be monitored by ^{31}P MRS, changes in the PME peak that can also be monitored by ^{31}P MRS, and tCho compound changes, which

Using Metabolomics to Monitor Anticancer Drugs

can be monitored by in vivo ^{1}H MRS. Not all of the responses noted here were actually measured in solid tumours in vivo. Some were measured in extracts of cultured cells and others by ex vivo NMR of tumour biopsies, but in principle it should be possible to monitor such drug actions by noninvasive MRS using currently available 1.5T and 3T instruments, at least if the tumours are in superficial bodily sites such as the lymph nodes, the breast, etc. Thus, this limited subset of metabolomic features (bioenergetic metabolites, phosphomonoesters and choline compounds), which show characteristic changes in response to anticancer drugs, could be monitored in some clinical trials. In the brain, it is possible to monitor several more metabolites by ^{1}H MRS (lactate, N-acetylaspartate, creatine and myo-inositol) so more sophisticated studies might be possible; however, very few anticancer drugs cross the blood–brain barrier (Murphy et al. 2004), so there is at present little need for a way to monitor chemotherapy in brain cancer. Tumours in the pelvis—particularly the prostate (Heerschap et al. 1997; Zakian et al. 2003) and also the cervix (Mahon et al. 2004a, b)—also give quite good ^{1}H MRS spectra.

Certain common features can be seen in the responses tabulated in Table 1. Increased bioenergetic metabolites were observed when tumours in animals were treated with the antimetabolite 5-FU and the alkylating agents ifosfamide and cyclophosphamide, indicating an apparently paradoxical improvement in the tumour's bioenergetic status. This phenomenon, which is not usually observed when tumours in patients (rather than animals) are treated, may perhaps be due to improved blood flow to the remaining tumour tissue as the tumour shrinks rapidly in response to the high drug doses that can be given to animals. Another factor may be the poor blood supply seen in the implanted subcutaneous tumours that are usually studied in animals. In contrast, drugs of several classes have been seen to cause decreases in tumour bioenergetic metabolites: a cyclin kinase inhibitor, an HSP90 inhibitor, a histone deacetylase inhibitor, two VDAs and two PI3 kinase inhibitors. The principle mechanism of action of the VDAs is obvious: they destroy the blood vessels and thus block the access of oxygen and nutrients. LAQ824 was the only one of the three histone deacetylase inhibitors to cause a decrease in bioenergetic metabolites, and this drug is known to have a vascular disruption action as well as inhibition of histone deacetylase (Qian et al. 2004). The mechanism(s) by which the other drugs reduce bioenergetic metabolites is currently unclear.

The drugs that caused increases or decreases in the PME peak in response to drug treatment are shown in the next column. Increased PME was observed in response to the antimetabolite 5-FU and the alkylating agents ifosfamide and cyclophosphamide, the HSP90 inhibitor 17AAG and the histone deacetylase inhibitors LAQ824 and SAHA. In contrast, decreased PME was observed in response to the choline kinase inhibitor MN58b and the antimicrotubule agent docetaxel.

The next column shows the drugs that caused increases and decreases in the tCho signal following drug therapy. Increased tCho was observed in response to the histone deacetylase inhibitor SAHA. Reduced tCho was found following treatment with the choline kinase inhibitor MN58b; the HIF-1α inhibitor PX478, and two of the vascular disrupting agents DMXAA and ZD6126.

The penultimate column lists the changes in phospholipid metabolites (i.e. PC, PE, GPC and GPE) and free choline following therapy, assessed by ex vivo ^{1}H or ^{31}P NMR. The changes of these metabolites confirmed the in vivo modulation of PME and/or PDE (by in vivo ^{31}P MRS) and tCho (by in vivo ^{1}H MRS). A reduced PC level was found following treatment with many different classes of anticancer drug. These drugs include the alkylating agent cyclophosphamide, the choline kinase inhibitor MN58b, the VDA ZD6126, the HIF-1α inhibitor PX478, the PI3K inhibitors—PI103, LY294002 and wortmannin—the MAPK inhibitor U0126, the fatty acid synthase inhibitor orlistat, and the antimicrotubule agent docetaxel. In cases where both in vivo and ex vivo measurements were carried out, a drop in PC is associated with decreased PME and/or tCho. Reduction in the PC level is generally associated with decreased cell membrane synthesis and proliferation, which is consistent with the expected response following therapy (Ackerstaff et al. 2003; de Certaines et al. 1993).

PME increases due to an elevated level of PE were found in response to the alkylating agent ifosfamide. A rise in the PE/PC ratio (due to a drop in PC) was seen in response to the antimetabolite 5-FU and the alkylating agent cyclophosphamide. However, increases in PME and/or tCho were also observed following treatment with the HSP90 inhibitor 17AAG and two of the histone deacetylase inhibitors, LAQ824 and

Table 1 Summary of responses to drugs that can be detected by ^{31}P or ^{1}H MRS in vivo and ex vivo. In some cases, the assays reported were performed ex vivo, but in principle the metabolites in question could be detected noninvasively in vivo. Unless otherwise stated, measurements were performed on tumours in animals

	Drug class	Bioenergetic metabolites (In vivo ^{31}P MRS)	PME (In vivo ^{31}P MRS)	tCho (In vivo ^{1}H MRS)	Phospholipids in the tCho region (Ex vivo ^{1}H or ^{31}P NMR)	References
5FU	Anti-metabolite	↑NTP/Pi, ↑PCr/Pi	↑PE/PC		↑PE, ↑GPC, ↑GPE	Street et al. 1997
Ifosfamide	Alkylating agent	↑β-ATP/Pi	↑PME/Pi		↑PE	Vaidya et al. 2003
Cyclophosphamide	Alkylating agent	↑NTP/Pi, ↑PCr/Pi	↑PE/PC		↓PC, ↑GPC, ↑GPE	Street et al. 1995
CYC202 (R-roscovitine)	Cyclin kinase inhibitor	↓NTP/TotP, ↑Pi/TotP, ↑Pi/NTP			↓GPC	Chung et al. 2002
17AAG	HSP90 inhibitor	↓β-NTP/TotP	↑PME/PDE, ↑PME/TotP, ↑PME/β-NTP		↑PC[a], ↑GPC[a], ↑PC, ↑PE	Chung et al. 2003
MN58b	Choline kinase inhibitor		↓PME/TotP	↓tCho	↓PC[a], ↓PC	Al-Saffar et al. 2006

Table 1 (continued)

	Drug class	Bioenergetic metabolites (In vivo ^{31}P MRS)	PME (In vivo ^{31}P MRS)	tCho (In vivo ^{1}H MRS)	Phospholipids in the tCho region (Ex vivo ^{1}H or ^{31}P NMR)	References
LAQ824	Histone deacetylase inhibitor	↓PCr/TotP ↓β-NTP/TotP ↓β-NTP/Pi ↑Pi/TotP	↑PME/TotP		↑PC, ↑PE, ↑Free choline, ↓GPC, ↓GPE	Chung et al. 2007
SAHA	Histone deacetylase inhibitor			↑tCho/totSignal[b] ↑tCho/totLipid[b]	↑PC[a]	Sankaranarayanapillai et al. 2005, 2007b
Phenyl-butyrate	Histone deacetylase inhibitor				↑GPC[a] ↑tCho[a]	Milkevitch et al. 2005
DMXAA	Vascular disrupting agent	↓β-NTP/Pi ↓PDE/PME		↓tCho	↓GPC, ↓GPE, ↑Free choline	McPhail et al. 2005
ZD6126	Vascular disrupting agent			↓tCho	↓PC, ↓GPC	Madhu et al. 2006
Combre-tastatin	Vascular	↓β-NTP/Pi[b]				Maxwell et al. 1998; Beauregard et al. 2001

Table 1 (continued)

	Drug class	Bioenergetic metabolites (In vivo ^{31}P MRS)	PME (In vivo ^{31}P MRS)	tCho (In vivo ^{1}H MRS)	Phospholipids in the tCho region (Ex vivo ^{1}H or ^{31}P NMR)	References
A4 Phosphate	disrupting agent	↑Pi/NTP				
PX478	HIF-1α inhibitor			↓tCho	↓PC, ↓PE, ↓GPC, ↓GPE	Jordan et al. 2005
LY294002 and Wortmannin	PI3K inhibitors	↓NTP[a] (by ex vivo ^{31}P NMR)			↓PC[a] ↑GPC[a]	Beloueche-Babari et al. 2006
PI103	PI3K inhibitor	↓NTP[a] (by ex vivo ^{31}P NMR)			↓PC[a] ↓PE[a]	Al-Saffar et al. 2007
U0126	MAPK inhibitor				↓PC[a]	Beloueche-Babari et al. 2005
Orlistate	Fatty acid synthase inhibitor				↓PC[a]	Ross et al. 2007
Docetaxel	Antimicro-tubule agent		↓PC		↓PC, ↑GPC	Morse et al. 2007

[a]Measured in cultured cells
[b]See text for more details

SAHA. In these cases the PME change was due to increases in PC and PE or PC alone. A rise in PC following a positive response to an anticancer drug treatment is unusual, as elevated PC is normally associated with tumour growth (Ackerstaff et al. 2003; Podo 1999; Aboagye and Bhujwalla 1999). A rise in PC was found in response to the Hsp90 inhibitor and two of the HDAC inhibitors; these two classes of drug both exert inhibitory effects on HSP90 but the mechanism behind this PC increase remained unclear. However, Sreedhar et al. (2003) reported that HSP90 plays an important role in the maintenance of cellular integrity. Hence, one might speculate that the rise in PC may be caused by the release of PC from the cell membrane following the compromised cellular integrity due to inhibition of HSP90.

GPC and GPE are associated with cell membrane breakdown. Changes of GPC and GPE are found to occur following responses to many different classes of anticancer drug but the mechanisms behind these changes remain unclear and require further investigation.

^{31}P MRS studies have been conducted for many years on the effect of classical anticancer drugs on tumours in vivo (reviewed by Negendank 1992; de Certaines et al. 1993). In general, it has been found that the PME signals are the most useful for monitoring drug treatments by ^{31}P MRS, as they fall in response to most forms of therapy. The measurements of PME reported in these two earlier reviews were mainly performed by ^{31}P MRS in vivo. This method has the advantage that it is noninvasive, but in most cases only a single PME peak can be resolved in vivo. In contrast, the studies reviewed in the present work include many measurements performed using ^1H MRS in vivo and ^1H and/or ^{31}P NMR ex vivo on extracts of cultured cells and tumour biopsies, or by HR-MAS NMR on solid tumour biopsies. In recent years, in vivo ^1H MRS to measure tCho for monitoring drug therapy became more widely used and a fall in tCho has been found following a number of types of drug treatment. The ex vivo methods that are also reported in the present review resolve more metabolites because of improved spectral resolution. They are complementary to the in vivo measurements, as they can provide additional information and help in the interpretation of the in vivo data. For instance, ex vivo methods can be used to pinpoint the mechanisms underlying the modulation of the PME or tCho signals.

Using Metabolomics to Monitor Anticancer Drugs 75

This review has demonstrated that metabolomics by MRS and NMR methods has many applications for monitoring pharmacodynamics of novel anticancer drugs. Much work remains to be done, however, on the metabolic mechanisms underlying the effects observed.

References

Aboagye EO, Bhujwalla ZM (1999) Malignant transformation alters membrane choline phospholipid metabolism of human mammary epithelial cells. Cancer Res 59:80–84

Ackerstaff E, Glunde K, Bhujwalla ZM (2003) Choline phospholipid metabolism: a target in cancer cells? J Cellular Biochem 90:525–533

Al-Saffar NMS, Troy H, Ramirez de Molina A, Jackson LE, Madhu B, Griffiths JR, Leach MO, Workman P, Lacal JC, Judson IR, Chung Y-L (2006) Noninvasive magnetic resonance spectroscopic pharmacodynamic markers of the choline kinase inhibitor MN58b in human carcinoma models. Cancer Res 66:427–434

Al-Saffar NMS, Jackson LE, Raynaud F, de Molina AM, Lacal JC, Workman P, Leach MO (2007) PI3K inhibition using a novel inhibitor deregulates choline kinase resulting in PC depletion detected by MRS. Proceedings of ISMRM, pp. 125

Beauregard DA, Hill SA, Chaplin DJ, Brindle KM (2001) The susceptibility of tumors to the antivascular drug combretastatin A4 phosphate correlates with vascular permeability. Cancer Res 61:6811–6815

Beloueche-Babari M, Jackson LE, Al-Saffar NMS, Workman P, Leach MO, Ronen SM (2005) Magnetic resonance spectroscopy monitoring of mitogen-activated protein kinase signaling inhibition. Cancer Res 65:3356–3363

Beloueche-Babari M, Jackson LE, Al-Saffar NMS, Eccles SA, Raynaud FI, Workman P, Leach MO, Ronen SM (2006) Identification of magnetic resonance detectable metabolic changes associated with inhibition of phosphoinositide 3-kinase signaling in human breast cancer cells. Mol Cancer Ther 5:187–196

Chen AP, Grem JL (1992) Antimetabolites. Curr Opin Oncol 4:1089–1098

Chung Y-L, Troy H, Judson IR, Leach MO, Stubbs M, Ronen S, Workman P, Griffiths JR (2002) The effects of CYC202 on tumors monitored by magnetic resonance spectroscopy. Proceedings of AACR, pp. 1664

Chung Y-L, Troy H, Banerji U, Jackson LE, Walton MI, Stubbs M, Griffiths JR, Judson IR Leach MO, Workman P, Ronen SM (2003) Magnetic resonance spectroscopic pharmacodynamic markers of the heat shock protein 90 inhibitor 17-allylamino, 17-demethoxygeldanamycin (17AAG) in human colon cancer models. J Natl Cancer Inst 95:1624–1633

Chung Y-L, Troy H, Kristeleit R, Aherne W, Judson IR, Atadja P, Workman P, Leach MO, Griffiths JR (2007) MRS pharmacodynamic markers of a novel histone deacetylase inhibitor LAQ824, in a human colon carcinoma model. Proceedingc of ISMRM, pp. 2823

De Certaines JD, Larsen VA, Podo F, Carpinelli G, Briot O, Henriksen O (1993) In vivo 31P MRS of experimental tumours. NMR Biomed 6:345–365

Evanochko WT, Sakai TT, Ng TC, Krishna NR, Kim HD, Zeidler RB, Ghanta VK, Brockman RW, Schiffer LM, Braunschweiger PG et al. (1984) NMR study of in vivo RIF-1 tumors. Analysis of perchloric acid extracts and identification of 1H, 31P and 13C resonances. Biochim Biophys Acta 805:104–116

Govindaraju V, Young K, Maudsley A (2000) Proton NMR chemical shifts and coupling constants for brain metabolites. NMR Biomed 13:129–153

Heerschap A, Jager GJ, van der Graaf M, Barentsz JO, de la Rosette JJ, Oosterhof GO, Ruijter ET, Ruijs SH (1997) In vivo proton MR spectroscopy reveals altered metabolite content in malignant prostate tissue. Anticancer Res 17:1455–1460

Jordan BF, Black K, Robey IF, Runquist M, Powis G, Gillies RJ (2005) Metabolite changes in HT-29 xenograft tumors following HIF-1α inhibition with PX-478 as studied by MR spectroscopy in vivo and ex vivo. NMR Biomed 18:430–439

MacPhail LB, Chung Y-L, Madhu B, Clark S, Griffiths SR, Kelland LR, Robinson SP (2005) An investigation of tumor dose response to the vascular disrupting agent 5,6-dimethylxanthenone-4-acetic acid (DMXAA), using in vivo magnetic resonance spectroscopy. Clin Cancer Res 11:3705–3713

Madhu B, Waterton JC, Griffiths JR, Ryan AJ, Robinson SP (2006) The response of RIF-1 fibrosarcomas to the vascular-disrupting agent ZD6126 assessed by in vivo and ex vivo 1H magnetic resonance spectroscopy. Neoplasia 8:560–567

Mahon MM, Cox IJ, Dina R, Soutter WP, McIndoe GA, Williams AD, deSouza NM (2004a) 1H magnetic resonance spectroscopy of preinvasive and invasive cervical cancer: in vivo-ex vivo profiles and effect of tumor load. J Magn Reson Imaging 19:356–364

Mahon MM, Williams AD, Soutter WP, Cox IJ, McIndoe GA, Coutts GA, Dina R, deSouza NM (2004b) 1H magnetic resonance spectroscopy of invasive cervical cancer: an in vivo study with ex vivo corroboration. NMR Biomed 17:1–9

Maxwell RJ, Nielsen FU, Breidahl T, Stødkilde-Jørgensen H, Horsman MR (1998) Effects of combretastatin on murine tumours monitored by 31P MRS, 1H MRS and 1H MRI. Int J Radiat Oncol Biol Phys 42:891–894

Milkevitch M, Shim H, Pilatus U, Pickup S, Wehrle JP, Samid D, Poptani H, Glickson JD, Delikatny EJ (2005) Increases in NMR-visible lipid and glycerophosphocholine during phenylbutyrate-induced apoptosis in human prostate cancer cells. Biochim Biophys Acta 1734:1–12

Morse DL, Raghunand N, Sadarangani P, Murthi S, Job C, Day S, Howison C, Gillies RJ (2007) Response of choline metabolites to docetaxel therapy is quantified in vivo by localized ^{31}P MRS of human breast cancer xenografts and in vitro by high-resolution ^{31}P NMR spectroscopy of cell extracts. Magn Reson Med 58:270–280

Murphy PS, Viviers L, Abson C, Rowland IJ, Brada M, Leach MO, Dzik-Jurasz AS (2004) Monitoring temozolomide treatment of low-grade glioma with proton magnetic resonance spectroscopy. Br J Cancer 90:781–786

Neckers L (2002) Hsp90 inhibitions as novel cancer chemotherapeutic agents. Trends Mol Med 8:S55–S60

Negendank WG (1992) Studies of human tumors by MRS: a review. NMR Biomed 5:303–324

Nelson C, Moffat B, Jacobsen N, Henzel WJ, Stults JT, King KL, McMurtrey A, Vandlen R, Spencer SA (1996) Glycerophosphoethanolamine (GPEA) identified as an hepatocyte growth stimulator in liver extracts. Exp Cell Res 229:20–26

Podo F (1999) Tumor phospholipid metabolism. NMR Biomed 12:413–439

Qian DZ, Wang X, Kachhap SK, Kato Y, Wei Y, Zhang L, Atadja P, Pili R (2004) The histone deacetylase inhibitor NVP-LAQ824 inhibits angiogenesis and has a greater antitumor effect in combination with the vascular endothelial growth factor reception tyrosine kinase inhibitor PTK787/ZK222584. Cancer Res 64:6626–6634

Ross J, Tong WP, Kaluarachi K, Ronen SM (2007) Detection of metabolic effects of fatty acid synthase inhibition by magnetic resonance spectroscopy. Proceedings of ISMRM, pp. 121

Sankaranarayanapillai M, Tong WP, Maxwell DA, Pal A, Pang J, Bornmann WG, Gelovani JG, Ronen SM (2006) Detection of histone deacetylase inhibition by noninvasive magnetic resonance spectroscopy. Mol Cancer Ther 5:1325–1334

Sankaranarayanapillai M, Kaluarachchi K, Ronen SM (2007a) 13C MRS detection of increased choline metabolism following HDAC inhibition. Proceedings of ISMRM, pp. 123

Sankaranarayanapillai M, Bankson JA, Yuan Q, Dafni H, Webb D, Pal A, Jackson EF, Gelovani J, Tong WP, Ronen SM (2007b) In vivo detection of histone deacetylase inhibition by MRS. Proceedings of ISMRM, pp. 2990

Shungu DC, Bhujwalla ZM, Wehrle JP, Glickson JD (1992) 1H NMR spectroscopy of subcutaneous tumors in mice: preliminary studies of effects of growth, chemotherapy and blood flow reduction. NMR Biomed 5:296–302

Sitter B, Sonnewald U, Spraul M, Fjosne HE, Gribbestad IS (2002) High-resolution magic angle spinning MRS of breast cancer tissue. NMR Biomed 15:327–337

Sreedhar AS, Mihaly K, Pato B, Schnaider T, Stetak A, Kis-Petik K, Fidy J, Simonics T, Maraz A, Csermely P (2003) Hsp90 inhibition accelerates cell lysis: anti-Hsp90 ribozyme reveals a complex mechanism of Hsp90 inhibitors involving both superoxide- and Hsp90-dependent events. J Bio Chem 278:35231–35240

Street JC, Mahmood U, Matei C, Koutcher JA (1995) In vivo and in vitro studies of cyclophosphamide chemotherapy in a mouse mammary carcinoma by 31P NMR spectroscopy. NMR Biomed 8:149–158

Street JC, Alfieri AA, Tragano F, Koutcher JA (1997) In vivo and ex vivo study of metabolic and cellular effects of 5-fluorouracil chemotherapy in a mouse mammary carcinoma. Magn Reson Imaging 15:587–596

Troy H, Chung Y-L, Judson IR, Leach MO, Stubbs M, Ronen S, Workman P, Griffiths JR (2002) The effects of the novel anticancer compound CYC202 on tumors monitored by magnetic resonance spectroscopy. Proceedings of ISMRM, pp. 2163

Vaidya S, Chung Y-L, Payne G, Leach M, Griffiths J, Pinkerton R (2003) Magnetic resonance spectroscopy studies of xenografted paediatric embryonal rhabdomyosarcoma. Br J Cancer 88 [Suppl 1]:517

Zakian KL, Eberhardt S, Hricak H, Shukla-Dave A, Kleinman S, Muruganandham M, Sircar K, Kattan MW, Reuter VE, Scardino PT, Koutcher JA (2003) Transition zone prostate cancer: metabolic characteristics at 1H MR spectroscopic imaging – initial results. Radiology 229:241–247

Ernst Schering Foundation Symposium Proceedings, Vol. 4, pp. 79–98
DOI 10.1007/2789_2008_090
© Springer-Verlag Berlin Heidelberg
Published Online: 06 June 2008

Biomarker Discovery for Drug Development and Translational Medicine Using Metabonomics

H.C. Keun(✉)

Biomolecular Medicine, Division of Surgery, Oncology, Reproductive Biology
and Anaesthetics, Imperial College Faculty of Medicine, Sir Alexander Fleming Building,
Exhibition Road, SW7 2AZ South Kensington, London, UK
email: *h.keun@imperial.ac.uk*

1	The Potential of the Metabolome to Fill the Biomarker Gap	80
2	Metabonomics in Toxicology	83
3	Metabonomics in Oncology	90
References		94

Abstract. There exists at present an urgent desire for better biomarkers, especially in the context of pharmaceutical drug development and in the detection and management of disease. Many researchers in the area of biomarker discovery and development have turned to the "-omics" sciences as a way of addressing these needs. Metabolic profiling, or metabonomics, defines the metabolic phenotype and offers a source of novel biomarkers that have better potential to translate effectively. This review will discuss the broad philosophy and motivations behind metabonomics, and illustrate the case with applications relevant to pharmaceutical development and patient management. Particular focus will be paid to the potential of metabonomics to contribute to biomarker discovery in toxicology and cancer research.

1 The Potential of the Metabolome to Fill the Biomarker Gap

The high candidate drug attrition rate, particularly late in the development process, is extremely costly and is often due to the poor translation/prediction of drug metabolism, efficacy or toxicity from preclinical models to humans. These problems are exacerbated and intertwined with the lack of mechanistic understanding relating to many disease processes and indeed even in terms of pharmacological action itself. Given that most disease is heterogenenous in phenotype, it is also widely recognized that there has been a dearth of biomarkers able to stratify patients into those groups most likely to benefit from a particular treatment, and that this has led to costly failures of development programmes (Frank and Hargreaves 2003). Such biomarkers are in fact a fundamental part of chemopreventative strategies, which require by default a means by which to select subpopulations at particularly high risk of disease or disease progression. Hence the need for biomarkers that can personalise medicine for the individual and that translate effectively between models and humans is paramount.

The establishment of platforms that can characterise a biological sample in an untargeted and highly parallel manner have revolutionised modern biological research. Operating in a primarily hypothesis-generating rather than hypothesis-testing mode allows for efficient screening of candidate biomarkers without making prior (and not always correct) assumptions about what relationships may be detected. Where it is possible to operate in a near comprehensive manner, such as in the definition of the genome or the associated transcriptional profile (transcriptome), one can be confident in obtaining a truly global perspective of a system at the chosen biomolecular level. By its nature "-omics" science is clearly technology driven, and its growth has only been possible by major and continuous advances in analytical science and bioinformatics. While this paradigm for research is somewhat challenging to the traditional reductionist approach to biology, it has begun to be routinely used by many investigators.

Analogous to the concept of the genome or proteome, the metabolome can be defined as the complete description of metabolite levels in a biological system (Tweeddale et al. 1999). Seen from a genocentric perspective, it offers a holistic description of the metabolic phenotype for functional genomics studies (Raamsdonk et al. 2001). From a more integrated viewpoint, knowledge of the metabolome, in addition to gene expression and regulation, is a vital component of systems biology. In reality, the metabolome is also subject to major exogenous inputs in the form of exposure to pharmaceutical, environmental and dietary compounds, and in most multicellular organisms is manipulated by means beyond the host genome or proteome via commensural microbes (Nicholson et al. 2004, 2005). Thus the metabolome is a diffuse concept and the researchers in this area (metabonomics/metabolomics) tend to work with data that can be more readily recognised as metabolic profiles: descriptions of small molecule composition that are not necessarily comprehensive but are largely unbiased in scope and amenable to quantitative interpretation. Analytical techniques used to measure metabolites, such as NMR spectroscopy and mass spectrometry, can be used to generate such profiles in a targeted or untargeted manner, efficiently defining detectable portions of the metabolome. While obviously desirable, metabolic profiles need not be fully resolved and annotated, i.e. all metabolites defining the profile are identified a priori). Analytical profiles can be analysed directly by statistical pattern recognition to identify factors that correlate to exposure to known toxicants or to the presence or likelihood of disease, and hence target metabolite characterisation. This distinction between targeted and untargeted metabolic profiling has some parallels with the difference between bottom-up versus top-down strategies for systems biology.

Among the ideas that surround metabolic profiling, the concept of metabonomics (Nicholson et al. 1999) in particular embraces a top-down approach, and traditionally has exploited the analysis of biofluids such as urine or plasma, lending it towards efforts to understand integrative physiological and systemic change (Fig. 1). Also, as part of metabonomic studies, intact tissue is often analysed by magic-angle spinning (MAS) NMR spectroscopy, which provides observations that have some relevance to *in vivo* NMR spectroscopy (MRS/MRI). The combination of biofluid analysis and untargeted metabolic analysis makes metabonomics an ideal platform for translational biomarker research. Noninvasive or minimally invasive biomarkers derived from body fluids or that can be detected by imaging are inherently more practical to take from

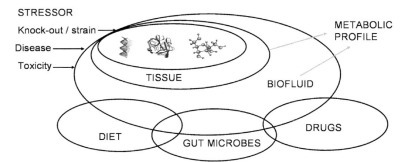

Fig. 1. Metabonomics works at the interface between an organism and its environment

the bench to the bedside. In addition, a metabolite is a defined chemical entity that is the same across all cell types, species and individuals, unlike gene products which change in sequence, splicing and are modified post-translation. This makes the analytical protocols used for metabolite detection fundamentally more likely to translate from models to humans and vice-versa.

Metabonomic profiles have been shown in principle to reflect the presence of pathological events in a number of disease models, including those for diabetes and metabolic syndrome (Dumas et al. 2007), infection (Wang et al. 2006), cancer (Al-Saffar et al. 2006; Bundy et al. 2006; Glunde et al. 2006; Teichert et al. 2008) and neurological disease (Tsang et al. 2006). Importantly, there are several key examples of metabonomics studies demonstrating the ability to detect the presence of, or potential for, disease in humans, namely atherosclerosis (Brindle et al. 2002; Makinen et al. 2007), cancer (Odunsi et al. 2005; Beger et al. 2006), schizophrenia (Holmes et al. 2006) and congenital defects in metabolism in infants, including those of unknown aetiology (Wevers et al. 1999). While the role of metabolism in the aetiology of disease processes such as metabolic syndrome is obvious, there are metabolic phenotypes for other pathologies such as neurological disease where the link is less clear, highlighting the potential of metabonomics to provide novel biological insight into already well-studied areas. Much of this new biology arises from the fact that metabolic profiles will not only be

Metabolic Profiling for Drug Development

determined by altered regulation of metabolic pathways and enzymatic activity, but also reflect environmental exposures and the functional integrity of cells, and tissues. Nongenomic factors already form the basis of most systemic biomarkers currently used in pathological reporting, particularly in toxicology.

2 Metabonomics in Toxicology

In toxicology, there are several ways in which new biomarkers can make an impact:

- By detecting otherwise silent pathologies;
- By being more translatable/relevant to humans;
- By predicting the individual susceptibility to an adverse event;
- By being less invasive and allowing response dynamics to be defined using less compound and fewer animals;
- By predicting traditional outcomes (i.e. acting as surrogate endpoints) and thus allowing risk and safety margins to be evaluated using less compound and fewer animals earlier in development;
- By revealing the mechanism leading to toxicity or the potential for toxicity and thus aid risk assessment.

There is a wealth of data in preclinical models demonstrating how the relationships between metabolic profiles and the severity, timing, site and mechanism of chemical toxicity could be exploited for all of these purposes (Robertson 2005; Keun 2006; Keun and Athersuch 2007).

Metabolic profiling can add significant value to the samples routinely generated by preclinical studies in drug discovery and development. In the context of such studies, endogenous metabolites are largely seen as interferences to the study of drug metabolites or other biomarkers of exposure. However, a substantial body of work has demonstrated that specific urinary metabolite changes could be associated with liver toxicity (Nicholson et al. 2002). Using model compounds, a number of particular biomarkers have been reported in metabonomic studies, including taurine for general liver dysfunction (Sanins et al. 1990); bile aciduria for billiary toxins (Robertson et al. 2000); N-methyl nicotinamide for peroxisome proliferation (Connor et al. 2004), 5-oxoprolinuria for disruption to glutathione metabolism (Waters et al. 2006) and medium chain dicarboxylic aciduria for dysfunction of mitochondrial fatty acid metabolism (Mortishire-Smith et al. 2004). A combination of markers was also shown to give site-specific information with regard to nephropathy and was sensitive to the severity and recovery of the lesion (Gartland et al. 1990; Holmes et al. 1992; Anthony et al. 1994).

Information derived from metabonomics could provide toxicological input in lead selection and optimisation where the amount of compound available is relatively low. By using urine that might already be collected for metabolism studies and by sampling continuously, early hazards can be detected efficiently, i.e. without extended dosing using extra animals. Using a multivariate regression model, five structurally similar compounds could be ranked based on NMR urinalysis, revealing a specific interruption of renal choline uptake that was not detected using classical methods of assessing toxicity (Dieterle et al. 2006). In this instance, metabonomics was able to detect the potential for an adverse event prior to the appearance of histopathology as well as provide clues as to the mechanism of toxicity, thus aiding risk assessment. All else being equal, the compounds producing a normal metabolic profile could be the better candidates for further development.

An important factor in interpreting changes in metabolic profile is the time course or trajectory. Urine sampling is noninvasive and effectively allows continuous monitoring over time. As variation in the dynamics of these metabolic perturbations also coincide with variation in the rate and severity of toxicity between individual animals, metabolite trajectories are important for understanding the specificity of a biomarker response (Nicholson et al. 2002). In principle, even a single molecule could be affected by several processes throughout an experiment, such as an adaptive or stress response, the loss of function or compartmentation, or significantly, regeneration (Fig. 2). It is difficult to correlate such changes directly to other endpoints that may be undersampled or have completely different time courses such as histopathology or gene expression (Fig. 3). We may wish to filter out the other factors and focus on the adaptive changes that might be the most relevant for predicting the chronic outcome from an acute study. One way we can tackle the

Metabolic Profiling for Drug Development 85

problem is to look at the relationship between the time courses of several metabolites. Metabolites that share the same trajectories presumably reflect the same underlying process, and understanding the collection of metabolites that respond together can help to refine the definition of these processes. For example, it has frequently been observed that the excretion pattern of hippurate and the Krebs cycle intermediates are frequently coincident in toxicological studies (Fig. 4). While several explanations of this phenomenon could exist for any given treatment, such as a mitochondrial specific response, it is known that all these molecules are similarly reduced by reduced food intake (Connor et al. 2004). We can then use this correlation of metabolite excretion to infer the process occurring, even when the magnitude of the effect is different from study to study. We can also use the pattern of correlations as a model from which to discern deviation from the influence of these processes and begin to attribute further biological significance to metabolite changes.

Many of these key ideas were developed within the Consortium for Metabonomic Toxicology (COMET) project (Lindon et al. 2003, 2005), which during its initial phase between 2001 and 2004 generated a database of over 35,000 biofluid metabolic profiles from 147 exposures in rodents to toxicological and physiological stressors. These data mostly included acute exposures to a wide variety of liver and kidney toxicants, but physiological stressors such as food restriction and partial hepatectomy were also examined, as was toxicity at other target organs such as testicular and pancreatic toxicity. This project was able to demonstrate that metabonomic responses to toxicity were (a) robust analytically and biologically using a multi-centre approach and high-throughput profiling (Keun et al. 2002, 2004) and (b) were sufficiently specific to the site and mechanism of toxicity to allow detection and classification of adverse events using a statistical model alone, the COMET expert system (Ebbels et al. 2007).

A large element of the expert system was the efficient handling of multivariate data via pattern recognition techniques. It had been shown previously that techniques such as principal components analysis (PCA) were potentially very valuable in visualisation and classification of metabonomics data (Holmes et al. 1992; Keun et al. 2004). Within this approach is the implicit assumption that allegedly similar profiles represent similar states and hence the same responses to toxin expo-

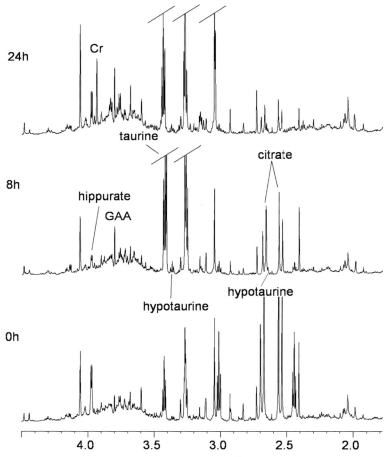

Fig. 2. Early effects to the aliphatic region of the ^{1}H NMR spectrum of rat urine after partial hepatectomy, a model of liver regeneration

sure, i.e. a compendium approach to toxicity classification. Putting this idea into practice across many studies required that highly multivariate metabolic trajectories be modelled. PCA allows trajectories to be visualised not in just one or two dimensions but using an infinite number

Metabolic Profiling for Drug Development

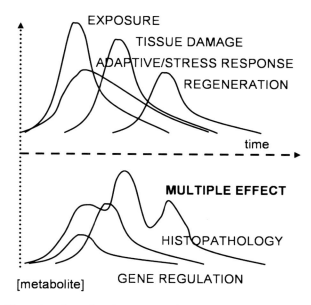

Fig. 3. Biomarker dynamics after acute toxin exposure

of measurements, summarising the variation into a lower dimensional space. It was subsequently shown that, after appropriate adjustment for the severity of response and the baseline metabolic profile, coincident metabolic trajectories result from the appearance of the same lesion via different chemical compounds (Keun et al. 2004). This led us to the homothetic trajectories hypothesis: it is the shape of the trajectory, i.e. how metabolite changes correlate to each other, that encapsulates the metabolic response in a manner best suited to classifying the toxicity.

For the COMET expert system, it was also necessary to compare trajectories easily and objectively. A nonlinear density estimation approach called CLOUDS was used (Ebbels et al. 2003). Related to Parzen density estimation, CLOUDS allows toxin-likeness to be assessed by superposition of trajectories. The overlap integral generated indicated the similarity of response while also taking into account the variability in response. In a training set of urinary NMR data from 80 studies, the

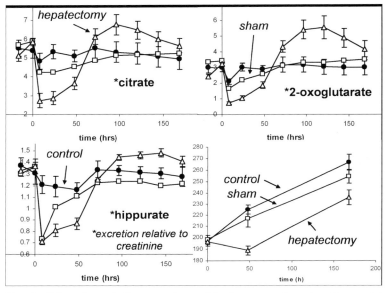

Fig. 4. Selected metabolite excretion and body mass trajectories after partial hepatectomy

approach was show to be able to cluster treatments according to target organ and even sub-organ specificity (Ebbels et al. 2007). In a predictive analysis it was possible to assign the correct target organ to the majority of treatments with 92% accuracy.

In a shift from observational to mechanistic application of metabonomics in toxicology the second phase of the COMET project will attempt to (a) establish biomarkers for renal papillary necrosis, an otherwise silent lesion and (b) define factors that contribute to the hypervariability of response to galactosamine (galN), a model for idiosyncratic hepatic toxicity. Metabonomic studies are already providing new insight into the protective role of glycine in galN toxicity, providing an example of the mechanistic role of the platform. In a ^{1}H NMR spectroscopic study, the level of N-acetylglucosamine (glcNAc) in the post-dose urine was found to correlate strongly with the degree of galN-induced liver

Metabolic Profiling for Drug Development

damage, while the urinary level of glcNAc was not significantly elevated in rats treated with both galN and glycine (Coen et al. 2007). Treatment with glycine alone was found to significantly increase hepatic levels of uridine, UDP-glucose and UDP-galactose. Uridine is also protective to galN toxicity, suggesting that the protective role of glycine against galN toxicity might be mediated by changes in the uridine nucleotide pool rather than by preventing Kupffer cell activation, as currently presumed.

In addition to correlating to the presence of toxicity, there is also evidence that metabolic profiles can directly predict the susceptibility of individual organisms to pathological events following toxicant exposure. Proof-of-principle studies have shown that the severity of drug hepatoxicity and drug metabolism in rodents can be predicted from pretreatment urinary metabolic profiles (Clayton et al. 2006), and a validation of the latter of these observations in a clinical trial is currently underway. In light of the analogous concept of pharmacogenetics, this model has been described as pharmaco-metabonomics. Interestingly, the prognostic urinary metabolites identified appeared to be diet-related compounds generated by commensural gut microbes, highlighting how metabonomics provides a unique viewpoint on extra-genomic interactions.

As metabolites from exogenous and endogenous sources are measured simultaneously in an unbiased manner, metabonomics is generally well suited to the generation of novel predictive biomarkers that link environmental exposures to human health via a meet-in-the-middle approach (Vineis and Perera 2007). The risk of developing cancer is clearly linked to dietary exposure to carcinogens, such as in the meat-derived heterocyclic amines (Gooderham et al. 2006), and exposure to chemopreventative agents, such as resveratrol found in red grapes (Aziz et al. 2003). Many more factors, either directly active toxicants or modulating agents could be discovered by a metabonomic approach, which will be a more common part of prospective biomarker studies, both epidemiological and clinical.

3   Metabonomics in Oncology

Irrespective of biomarker discovery, there is strong evidence for a common metabolic phenotype associated with cancer. As long ago as the 1920s, Otto Warburg described the phenomenon of aerobic glycolysis, the apparently greater tendency of tumour cells to convert glucose to lactate in the presence of normal oxygen conditions. Evidence exists to suggest that the glycolytic phenotype confers selective growth advantages to transformed cells (Gatenby and Gillies 2004) and the function of the tumour suppressor p53 has been linked to this phenomenon (Matoba et al. 2006). Other aspects of the tumour metabolic phenotype centre around the observation that growth of tumour cells in culture is often unusually dependent on the availability of common substrates, such as glutamine, methionine, cysteine and arginine (Wheatley 2005). In oncology, the altered metabolic phenotype of tumours is routinely exploited in diagnosis (e.g. FDG-PET) and in therapy (e.g. 5-FU, an antimetabolite). Metabolic profiling offers a number of opportunities for discovery and development of noninvasive biomarkers in cancer studies based on both screening and functional genomics strategies.

Metabolic profiles have been shown to be able to subclassify cancer phenotypes in number of solid tumours, including those of the brain (Tate et al. 2006), prostate (Cheng et al. 2005), ovary (Denkert et al. 2006) and breast (Katz-Brull et al. 2002). A key pathway involved in this discrimination is choline metabolism. The ubiquitous presence of elevated choline metabolites in tumour cells (Fig. 5) has been detected by magnetic resonance both *in vivo* and *in vitro*, translating across species and present across a wide range of primary and secondary tumour sites (Glunde et al. 2006). This, together with an increase in the phosphocholine/glycerophosphocholine (PC/GPC) ratio, has been shown to be a general marker for rapid proliferation and tumorigenicity, but in conjunction with other metabolite measurements could be predictive of the invasiveness of a tumour and useful in the clinical staging of disease. While effort continues to be invested in determining the mechanism behind this phenomenon, it would appear to be in part due to increased choline transport into the cell and upregulated choline kinase activity in response to the demand for phosphatidylcholine and membrane synthesis (Glunde et al. 2004, 2005). The observation has proved

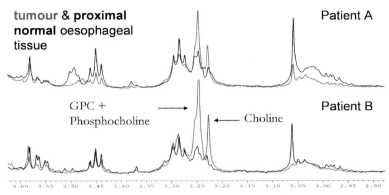

Fig. 5. Perturbations to choline metabolism detected by ^{1}H MAS-NMR spectroscopy of intact tissue

of particular value in the development of choline kinase inhibition as a therapeutic strategy, which has been shown to be successful in both HT29 and MDA-MB-231 xenograft models (Al-Saffar et al. 2006). This work tells an important story about the value of biomarker research via exploratory clinical studies in the translational research setting. Since the identification of choline kinase as a drug target has derived in part from the visible impact of its activity in tumours, we can immediately turn around the result and use the choline NMR signal as a pharmacodynamic marker, safe in the knowledge that it has clinical relevance (Fig. 6). It also provides a phenotypic anchor with which to help evaluate results in animal models.

We were interested in understanding how the metabolic phenotype of tumours in an autochonous model of prostate cancer (Transgenic Adenocarcinoma of Mouse Prostate; TRAMP) compared to the known human tumour profile. We found that while prostate-specific features of the human phenotype were preserved, such as a depletion of the unusually high levels of citrate in the prostate, the more general feature of elevated choline metabolites was not (Teichert et al., in press). Tumour tissue from the TRAMP model did not exhibit any upregulation of ChoK at either the transcriptional or protein level. These results helped to rationalise the lack of sensitivity of certain *in vivo* MRS parame-

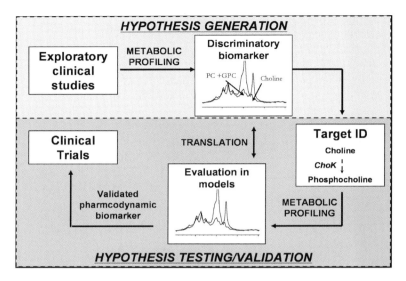

Fig. 6. The potential life cycle of a metabolic biomarker

ters such as the total choline/citrate ratio, which have known significance as a detection and progression marker in human disease, in the TRAMP model (Fricke et al. 2006). It is an example of how metabonomics can be used to evaluate preclinical models of cancer and suggests that the TRAMP model will behave differently pharmacodynamically to metabolically targeted therapies such as ChoK inhibition. These differences between the TRAMP tumour metabolic phenotype and the human disease may well originate from the specific form of oncogenic transformation (SV40 t & T antigen) used to generate the model that affects the products of p53 and Rb genes. It is interesting to note that the loss of the PTEN tumour suppressor, a more relevant event to human prostate cancer than the consequences of SV40 transfection, produces the expected metabolic response *in vivo* MRS data to suggest that it may be a more appropriate model for biomarker studies (Fricke et al. 2006). Such experiments do not only support biomarker development, but simultaneously further our understanding of the link between metabolism and malignant transformation.

Metabolic Profiling for Drug Development

To what extent the metabolic phenotype of cancer is causal or consequential to carcinogenesis and disease progression is still widely debated; however, key examples such as the pseudohypoxia effect of high succinate and fumarate levels in HIF activation show how metabolism can directly promote tumour development (Pollard et al. 2005). In light of this, metabolic profiling has clear potential to reveal new relationships between metabolic perturbation and transformation, and there is a good evidence base already available. Fibroblast cell lines progressively transformed from a primary to a cancerous state using telomerase, and oncogenic Ras show increasing sensitivity to glycolysis inhibitors while simultaneously becoming resistant to inhibition of oxidative metabolism. The choline metabolic phenotype is also mediated by both the loss of tumour suppressor function and oncogene activation, specifically p300 (Bundy et al. 2006), oncogenic Ras (Ratnam and Kent 1995), and telomerase (Iorio et al. 2005). While there are not yet clear examples of metabonomics predicting the future occurrence of cancer, there is *in vivo* evidence to support the hypothesis that some of the metabolic features of cancer can arise with premalignant transformation of cells. The presence of a biochemically abnormal field surrounding a tumour and either originating the neoplasm or caused by it can be defined by p53 mutation (Ito et al. 2005) and epigenetic changes (Ushijima 2007). It has been suggested that such a field could be detected by metabolic profiling, since correlations to the stage of disease were observed even in histologically normal tissue from patients with prostate cancer (Cheng et al. 2005).

Whether any of these effects manifest themselves in biofluids is not known, but clearly the discovery of systemic metabolic biomarkers specific to cancer is of enormous value in terms of patient management and detection screening. Although there is evidence that metabolic profiles of sera or serum lipids can detect the presence of ovarian (Odunsi et al. 2005) and pancreatic cancer (Beger et al. 2006), biomarker screening in serum or plasma has a difficult past with both NMR-based (Fossel et al. 1986; Okunieff et al. 1990) and SELDI-MS-based (Petricoin et al. 2002; Baggerly et al. 2005) protein marker profiles being challenged due to high normal variability and sample bias (Ransohoff 2005; Teahan et al. 2006). Thus despite the enormous potential of all "-omics" technologies, it is important to exercise some caution and to work hard to avoid historical pitfalls. Even in exploratory clinical studies, it is valuable and probably necessary to rationalise any putative metabolite biomarker firmly in the context of the tumour metabolic phenotype.

Acknowledgements. I would like to thank Mr. Danny Yakoub and Dr. Mary Bollard for their help in the production of the figures, and Miss Alexandra Backshall for proof-reading the manuscript.

References

Al-Saffar NMS, Troy H, Ramirez de Molina A et al. (2006) Noninvasive magnetic resonance spectroscopic pharmacodynamic markers of the choline kinase inhibitor MN58b in human carcinoma models. Cancer Res 66:427–434

Anthony ML, Sweatman BC, Beddell CR et al. (1994) Pattern-recognition classification of the site of nephrotoxicity based on metabolic data derived from proton nuclear-magnetic-resonance spectra of urine. Mol Pharmacol 46:199–211

Aziz MH, Kumar R, Ahmad N (2003) Cancer chemoprevention by resveratrol: in vitro and in vivo studies and the underlying mechanisms. Int J Oncol 23:17–28

Baggerly KA, Morris JS, Edmonson SR et al. (2005) Signal in noise: evaluating reported reproducibility of serum proteomic tests for ovarian cancer. J Natl Cancer Inst 97:307–309

Beger RD, Schnackenberg LK et al. (2006) Metabonomic models of human pancreatic cancer using 1D proton NMR spectra of lipids in plasma. Metabolomics 2:125–134

Brindle JT, Antti H, Holmes E et al. (2002) Rapid and noninvasive diagnosis of the presence and severity of coronary heart disease using H-1-NMR-based metabonomics. Nat Med 8:1439–1444

Bundy JG, Iyer NG, Gentile MS et al. (2006) Metabolic consequences of p300 gene deletion in human colon cancer cells. Cancer Res 66:7606–7614

Cheng LL, Burns MA, Taylor JL et al. (2005) Metabolic characterization of human prostate cancer with tissue magnetic resonance spectroscopy. Cancer Res 65:3030–3034

Clayton TA, Lindon JC, Cloarec O et al. (2006) Pharmaco-metabonomic phenotyping and personalized drug treatment. Nature 440:1073–1077

Coen M, Hong YS, Clayton TA et al. (2007) The mechanism of galactosamine toxicity revisited: a metabonomic study. J Proteome Res 6:2711–2719

Connor SC, Hodson MP, Ringeissen S et al. (2004a) Development of a multivariate statistical model to predict peroxisome proliferation in the rat, based on urinary H-1-NMR spectral patterns. Biomarkers 9:364–385

Connor SC, Wu W, Sweatman BC et al. (2004b) Effects of feeding and body weight loss on the H-1-NMR-based urine metabolic profiles of male Wistar Han rats: implications for biomarker discovery. Biomarkers 9:156–179

Denkert C, Budczies J, Kind T et al. (2006) Mass spectrometry-based metabolic profiling reveals different metabolite patterns in invasive ovarian carcinomas and ovarian borderline tumors. Cancer Res 66:10795–10804

Dieterle F, Schlotterbeck GT, Ross A et al. (2006) Application of metabonomics in a compound ranking study in early drug development revealing drug-induced excretion of choline into urine. Chem Res Toxicol 19:1175–1181

Dumas ME, Wilder SP, Bihoreau MT et al. (2007) Direct quantitative trait locus mapping of mammalian metabolic phenotypes in diabetic and normoglycemic rat models. Nat Genet 39:666–672

Ebbels T, Keun H et al. (2003) Toxicity classification from metabonomic data using a density superposition approach: 'CLOUDS'. Anal Chim Acta 490:109–122

Ebbels TMD, Keun HC, Beckonert OP et al. (2007) Prediction and classification of drug toxicity using probabilistic modeling of temporal metabolic data: the Consortium on Metabonomic Toxicology screening approach. J Proteome Res 6:4407–4422

Fossel ET, Carr JM, McDonagh J (1986) Detection of malignant tumors. Water-suppressed proton nuclear magnetic resonance spectroscopy of plasma. N Engl J Med 315:1369–1376

Frank R, Hargreaves R (2003) Clinical biomarkers in drug discovery and development. Nat Rev Drug Discov 2:566–580

Fricke ST, Rodriguez O, Vanmeter J et al. (2006) In vivo magnetic resonance volumetric and spectroscopic analysis of mouse prostate cancer models. Prostate 66:708–717

Gartland KPR, Anthony ML, Beddell CR et al. (1990) Proton Nmr-studies on the effects of uranyl-nitrate on the biochemical-composition of rat urine and plasma. J Pharmaceut Biomed Anal 8:951–954

Gatenby RA, Gillies RJ (2004) Why do cancers have high aerobic glycolysis? Nat Rev Cancer 4:891–899

Glunde K, Jie C, Bhujwall ZM (2004) Molecular causes of the aberrant choline phospholipid metabolism in breast cancer. Cancer Res 64:4270–4276

Glunde K, Raman V, Mori N et al. (2005) RNA interference-mediated choline kinase suppression in breast cancer cells induces differentiation and reduces proliferation. Cancer Res 65:11034–11043

Glunde K, Jacobs MA, Bhujwall ZM (2006) Choline metabolism in cancer: implications for diagnosis and therapy. Expert Rev Mol Diagn 6:821–829

Gooderham NJ, Lauber SN, Lauber SN et al. (2006) Mechanisms of action of carcinogenic heterocyclic amines. Toxicol Lett 164:S61–S62

Holmes E, Bonner FW, Sweatman BC et al. (1992) Nuclear-magnetic-resonance spectroscopy and pattern-recognition analysis of the biochemical processes associated with the progression of and recovery from nephrotoxic lesions in the rat induced by mercury(Ii) chloride and 2-bromoethanamine. Mol Pharmacol 42:922–930

Holmes E, Tsang TM, Huang JT et al. (2006) Metabolic profiling of CSF: evidence that early intervention may impact on disease progression and outcome in schizophrenia. Plos Medicine 3:e327

Iorio E, Mezzanzanica D, Alberti P et al. (2005) Alterations of choline phospholipid metabolism in ovarian tumor progression. Cancer Res 65:9369–9376

Ito S, Ohga T, Saeki H et al. (2005) p53 mutation profiling of multiple esophageal carcinoma using laser capture microdissection to demonstrate field carcinogenesis. Int J Cancer 113:22–28

Katz-Brull R, Lavin PT, Lenkinski RE (2002) Clinical utility of proton MRS in characterizing breast lesions. Radiology 225:650–651

Keun HC (2006) Metabonomic modeling of drug toxicity. Pharmacol Ther 109:92–106

Keun HC, Athersuch TJ (2007) Application of metabonomics in drug development. Pharmacogenomics 8:731–741

Keun HC, Ebbels TMD et al. (2002) Analytical reproducibility in H-1 NMR-based metabonomic urinalysis. Chem Res Toxicol 15:1380–1386

Keun HC, Ebbels TMD et al. (2004) Geometric trajectory analysis of metabolic responses to toxicity can define treatment specific profiles. Chem Res Toxicol 17:579–587

Lindon JC, Nicholson JK, Holmes E et al. (2003) Contemporary issues in toxicology – the role of metabonomics in toxicology and its evaluation by the COMET project. Toxicol Appl Pharmacol 187:137–146

Lindon JC, Keun HC, Ebbels TM et al (2005) The Consortium for Metabonomic Toxicology (COMET): aims, activities and achievements. Pharmacogenomics 6:691–699

Makinen VP, Soininen P et al. (2007) Cardiovascular risk factors in type 1 diabetes: a metabonomic study by 1H NMR spectroscopy of serum. Atherosclerosis Suppl 8:218–218

Matoba S, Kang JG, Patino WD et al. (2006) p53 regulates mitochondrial respiration. Science 312:1650–1653

Mortishire-Smith RJ, Skiles GL, Lawrence JW et al. (2004) Use of metabonomics to identify impaired fatty acid metabolism as the mechanism of a drug-induced toxicity. Chem Res Toxicol 17:165–173

Nicholson JK, Lindon JC, Holmes E (1999) 'Metabonomics': understanding the metabolic responses of living systems to pathophysiological stimuli via multivariate statistical analysis of biological NMR spectroscopic data. Xenobiotica 29:1181–1189

Nicholson JK, Connelly J, Lindon JC et al. (2002) Metabonomics: a platform for studying drug toxicity and gene function. Nat Rev Drug Discov 1:153–161

Nicholson JK, Holmes E, Lindon JC et al. (2004) The challenges of modeling mammalian biocomplexity. Nat Biotechnol 22:1268–1274

Nicholson JK, Holmes E, Lindon JC et al. (2005) Gut microorganisms, mammalian metabolism and personalized health care. Nat Rev Microbiol 3:431–438

Odunsi K, Wollman RM, Ambroson CB et al. (2005) Detection of epithelial ovarian cancer using H-1-NMR-based metabonomics. Int J Cancer 113:782–788

Okunieff P, Zietman A, Kahn J et al (1990) Lack of efficacy of water-suppressed proton nuclear-magnetic-resonance spectroscopy of plasma for the detection of malignant tumors. N Engl J Med 322:953–958

Petricoin EF, Ardekani AM, Hitt BA et al. (2002) Use of proteomic patterns in serum to identify ovarian cancer. Lancet 359:572–577

Pollard PJ, Briere JJ, Alam NA et al. (2005) Accumulation of Krebs cycle intermediates and over-expression of HIF1 alpha in tumours which result from germline FH, SDH mutations. Human Molr Genet 14:2231–2239

Raamsdonk LM, Teusink B, Broadhurst D et al. (2001) A functional genomics strategy that uses metabolome data to reveal the phenotype of silent mutations. Nat Biotechnol 19:45–50

Ransohoff DF (2005) Opinion – bias as a threat to the validity of cancer molecular-marker research. Nat Rev Cancer 5:142–149

Ratnam S, Kent C (1995) Early increase in choline kinase-activity upon induction of the H-Ras oncogene in mouse fibroblast cell lines. Arch Biochem Biophys 323:313–322

Robertson DG (2005) Metabonomics in toxicology: a review. Toxicol Sci 85:809–822

Robertson DG, Reily MD, Sigler RE et al. (2000) Metabonomics: evaluation of nuclear magnetic resonance (NMR) and pattern recognition technology for rapid in vivo screening of liver and kidney toxicants. Toxicol Sci 57:326–337

Sanins SM, Nicholson JK, Elcombe C et al. (1990) Hepatotoxin-induced hypertaurinuria – a proton Nmr study. Arch Toxicol 64:407–411

Tate AR, Underwood J, Acosta DM et al. (2006) Development of a decision support system for diagnosis and grading of brain tumours using in vivo magnetic resonance single voxel spectra. Nmr Biomed 19:411–434

Teahan O, Gamble S, Holmes E et al. (2006) Impact of analytical bias in metabonomic studies of human blood serum and plasma. Anal Chem 78:4307–4318

Teichert F, Verschoyle RD et al. (2008) Metabolic Profiling of Transgenic Adenocarcinoma of Mouse Prostate (TRAMP) tissue by 1H-NMR analysis – evidence for unusual phospholipid metabolism. Prostate

Tsang TA, Woodman B et al. (2006) Metabolic characterization of the R6/2 transgenic mouse model of Huntington's disease by high-resolution MAS H-1 NMR spectroscopy. J Proteome Res 5:483–492

Tweeddale H, Notley-McRobb L, Ferenci T (1999) Assessing the effect of reactive oxygen species on *Escherichia coli* using a metabolome approach. Redox Rep 4:237–241

Ushijima T (2007) Epigenetic field for cancerization. J Biochem Mol Biol 40:142–150

Vineis P, Perera F (2007) Molecular epidemiology and biomarkers in etiologic cancer research: the new in light of the old. Cancer Epidemiol Biomarkers Prev 16:1954–1965

Wang YL, Utzinger J, Xiao SH et al. (2006) System level metabolic effects of a *Schistosoma japonicum* infection in the Syrian hamster. Mol Biochem Parasitol 146:1–9

Waters NJ, Waterfield CJ, Farrant RD et al. (2006) Integrated metabonomic analysis of bromobenzene-induced hepatotoxicity: novel induction of 5-oxoprolinosis. J Proteome Res 5:1448–1459

Wevers RA, Engelke UFH et al. (1999) H-1-NMR spectroscopy of body fluids: inborn errors of purine and pyrimidine metabolism. Clin Chem 45:539–548

Wheatley DN (2005) Arginine deprivation and metabolomics: important aspects of intermediary metabolism in relation to the differential sensitivity of normal and tumour cells. Semin Cancer Biol 15:247–253

Ernst Schering Foundation Symposium Proceedings, Vol. 4, pp. 99–124
DOI 10.1007/2789_2008_091
© Springer-Verlag Berlin Heidelberg
Published Online: 25 June 2008

Pyruvate Kinase Type M2:
A Key Regulator Within the Tumour
Metabolome and a Tool for Metabolic
Profiling of Tumours

S. Mazurek[✉]

ScheBo Biotech AG, Netanyastrasse 3, 35394 Giessen, Germany
email: *s.mazurek@schebo.com, Sybille.Mazurek@vetmed.uni-giessen.de*

1	Introduction	100
2	The Pyruvate Kinase Isoenzymes	101
3	Bifunctional Role of the Pyruvate Kinase Isoenzyme Type M2 Within the Tumour Metabolome	103
4	Interaction of M2-PK with Different Oncoproteins	109
4.1	Interaction Between M2-PK and pp60v-src	109
4.2	Interaction Between M2-PK and A-Raf	110
4.3	Interaction Between M2-PK and Protein Kinase C Delta	112
4.4	Interaction Between M2-PK and HPV-16 E7	112
4.5	Interaction Between M2-PK and HERC1	114
5	Role of M2-PK in the Nucleus	115
6	Tumour M2-PK: A Biomarker for Metabolic Profiling of Tumours	115
7	Conclusions	118
	References	118

Abstract. Normal proliferating cells and tumour cells in particular express the pyruvate kinase isoenzyme type M2 (M2-PK, PKM2). The quaternary structure of M2-PK determines whether the glucose carbons are degraded to pyruvate and lactate with production of energy (tetrameric form) or channelled into synthetic processes, debranching from glycolytic intermediates such as nucleic acid, amino acid and phospholipid synthesis. The tetramer:dimer ratio of M2-

PK is regulated by metabolic intermediates, such as fructose 1,6-P2 and direct interaction with different oncoproteins, such as $pp60^{v\text{-}src}$ kinase, HPV-16 E7 and A-Raf. The metabolic function of the interaction between M2-PK and the HERC1 oncoprotein remains unknown. Thus, M2-PK is a meeting point for different oncogenes and metabolism. In tumour cells, the dimeric form of M2-PK is predominant and has therefore been termed Tumour M2-PK. Tumour M2-PK is released from tumours into the blood and from gastrointestinal tumours also into the stool of tumour patients. The quantification of Tumour M2-PK in EDTA plasma and stool is a tool for early detection of tumours and therapy control.

1   Introduction

The first oncogene discovered was the src oncogene. The discovery goes back to the year 1910 and Peyton Rous who found that cell-free cell extracts of sarcomas from Plymouth Rock hens transmit the disease when injected in other chickens (Rous 1910). Nearly contemporary to Peyton Rous, Otto Warburg began his investigations into the metabolism of tumour cells and in 1924 described for the first time that tumour cells produce high levels of lactate even in the presence of oxygen (Warburg et al. 1924). Both findings initiated two new fields of extensive and fundamental investigation. The basic observations of Peyton Rous resulted in the discovery of tumour viruses, oncogenes and proto-oncogenes. The transforming principle of the Rous sarcoma virus was isolated in 1977 and termed $pp60^{v\text{-}src}$ (Brugge and Erikson 1977). Three years later, it was demonstrated that $pp60^{v\text{-}src}$ is a protein tyrosine kinase, which was the first identified member of this class of enzymes (Hunter and Sefton 1980). In the field of metabolic research, it turned out that the glycolytic phenotype of tumour cells is the result of multiple mechanisms, which include activation of oncogenes as well as stabilization of transcription factors (Shim et al. 1997; Gatenby and Gillies 2004) and correlates with an upregulation of most glycolytic enzymes as well as changes in the isoenzyme composition of certain glycolytic enzymes (see Sect. 2). The tumour-specific metabolic phenotype is summarized as the tumour metabolome (http://www.metabolic-database.com). One of the glycolytic enzymes which was found to be consistently altered during tumorigenesis is pyruvate kinase. Tumour cells are characterized by the expression of the pyruvate kinase isoenzyme type M2 (M2-PK,

PKM2), which was found to be a target of the $pp60^{v\text{-}src}$ kinase (Presek et al. 1980, 1988) as well as other oncoproteins. Thus, M2-PK is a meeting point for different oncogenes and metabolism.

2 The Pyruvate Kinase Isoenzymes

In glycolysis, first two moles of ATP have to be invested in the hexokinase and 6-phosphofructo 1-kinase reaction before the ATP is regained in the phosphoglycerate kinase reaction. Net ATP production occurs in the last step within the glycolytic sequence, the dephosphorylation of phosphoenolpyruvate (PEP) to pyruvate catalyzed by pyruvate kinase (Fig. 1). In contrast to mitochondrial respiration, energy regeneration by pyruvate kinase is independent of oxygen and allows survival of the cells in the absence of oxygen, which is of special importance in tumour cells, often growing in areas with varying oxygen supply.

There are four pyruvate kinase isoenzymes in mammals which differ widely in their occurrence according to the type of tissue, kinetic characteristics and regulation mechanisms. The pyruvate kinase isoenzyme type L (L-PK) has the lowest affinity to its substrate PEP and is expressed in tissues with gluconeogenesis, such as liver, kidney and intestine (Brinck et al. 1994; Steinberg et al. 1999). L-PK is allosterically activated by fructose 1,6-P2 as well as ATP and phosphorylated by a cAMP-dependent protein kinase under the control of glucagon. The phosphorylation of L-PK leads to a reduction of the PEP affinity and an inactivation of the enzyme under physiological conditions. Furthermore, L-PK expression is regulated by nutrition. Whereas a carbohydrate-rich diet enhances protein synthesis of L-PK, hunger reduces L-PK expression.

The pyruvate kinase isoenzyme type M1 (M1-PK) has the highest affinity to its substrate PEP and is not allosterically regulated, phosphorylated or influenced by diet. M1-PK is expressed in skeletal muscle and brain, both organs which are strongly dependent upon a high rate of energy regeneration (Yamada and Noguchi 1999). The pyruvate kinase isoenzyme type R (R-PK) is found in erythrocytes and is very similar to L-PK in respect to its kinetic characteristics and regulation mechanisms (Noguchi et al. 1987). The pyruvate kinase isoenzyme type M2 (M2-

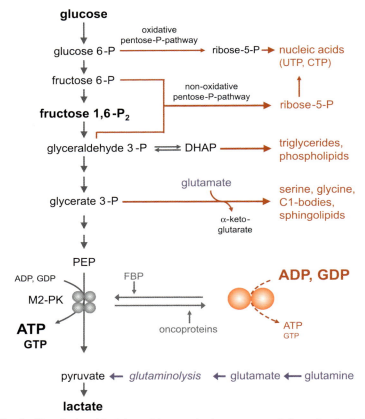

Fig. 1. Glycolysis with debranching synthetic processes. A large level of the highly active tetrameric form correlates with high levels of ATP and GTP, high ATP:ADP and GTP:GDP ratios as well as a high (ATP+GTP):(UTP+CTP) ratio. In contrast, high levels of the nearly inactive dimeric form of M2-PK correlate with low ATP and GTP levels, low ATP:ADP and GTP:GDP ratios as well as a low (ATP+GTP):(UTP+CTP) ratio

PK, PKM2) is expressed in some differentiated tissues, such as lung, fat tissue, retina as well as in all cells with a high rate of nucleic acid synthesis, which are all proliferating cells including normal proliferat-

M2-PK

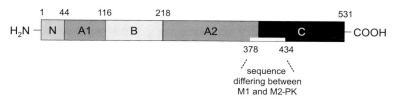

Fig. 2. Molecular structure of the M2-PK protein

ing cells, embryonic cells, stem cells and especially tumour cells. When embryonic cells differentiate, the M2-PK isoenzyme is progressively replaced by the respective tissue-specific pyruvate kinase isoenzymes. Conversely, during tumorigenesis the tissue-specific isoenzymes disappear and M2-PK is expressed (Reinacher and Eigenbrodt 1981; Staal and Rijksen 1991; Steinberg et al. 1999). The kinetic characteristics of the M2-PK isoenzyme depend on the quaternary structure of the enzyme (see Sect. 3).

The R and L isoenzymes of pyruvate kinase are encoded by the same gene and are expressed under the control of different tissue specific promotors (Noguchi et al. 1987). In the same way, the M1 and M2-PK isoenzymes are encoded by one gene but result from alternative splicing of exons 9 and 10. The human M1 and M2-PK isoenzyme differ in 23 amino acids within a stretch of 56 amino acids (Fig. 2) (Noguchi et al. 1986; Yamada and Noguchi 1999; Dombrauckas et al. 2005).

3 Bifunctional Role of the Pyruvate Kinase Isoenzyme Type M2 Within the Tumour Metabolome

The human M2-PK isoenzyme consists of 531 amino acids and can be subdivided into the N-terminal domain from aa 1–43, the A-domain which is composed of aa 44–116 as well as 219–389, the B-domain from aa 117–218 and the C-domain from aa 390–531 (Fig. 2) (Dombrauckas et al. 2005). The A-domain is responsible for the intermolecular subunit contact to compose a dimeric form. The interfaces of the

104 S. Mazurek

C-domain of two dimeric forms then associate to a tetrameric form. The C-domain contains 44 amino acids of the 56 amino acid stretch, which differs between the M1 and M2-PK subunits and is responsible for the different kinetic characteristics and regulation mechanisms found for M1 and M2-PK.

The upregulation of M2-PK is controlled by ras and the transcription factors SP1 and SP3 (Discher et al. 1998; Mazurek et al. 2001b). The M-gene furthermore has two HIF-binding sites (Kress et al. 1998; Stubbs et al. 2003; Brahimi-Horn and Pouyssegur 2007).

Whereas the other pyruvate kinase isoenzymes are characterized by a tetrameric quaternary structure, M2-PK may occur in a tetrameric form but also in a dimeric form. The tetrameric form of M2-PK has a high affinity to its substrate PEP and is highly active at physiological PEP concentrations. The dimeric form is characterized by a low affinity to PEP and is nearly inactive under physiological conditions (Fig. 3a).

Furthermore, the tetrameric form of M2-PK is associated with other glycolytic enzymes, such as hexokinase, glyceraldehyde 3-P dehydrogenase, phosphoglycerate kinase, enolase and lactate dehydrogenase, other enzymes such as nucleotide diphosphate kinase and adenylate kinase, components of the protein kinase cascade such as RAF, MEK and ERK as well as RNA in a cytosolic glycolytic enzyme complex (Hentze 1994; Nagy and Rigby 1995; Zwerschke et al. 1999; Mazurek et al. 2001a,b). The glycolytic enzyme complex can be isolated by isoelectric focusing using a buffer with low salt concentration. Proteins associated within the glycolytic enzyme complex focus at a common isoelectric point which is different than the isoelectric point of the purified proteins (Fig. 4). Migration of proteins in or out of the glycolytic enzyme complex are reflected by shifts in their isoelectric points. Accordingly, the dimeric form of M2-PK focuses outside the glycolytic enzyme complex at a more alkaline pH value. The close spatial proximity of the highly active tetrameric form of M2-PK to the other glycolytic enzymes of the

Fig. 3a,b. Affinity of the tetrameric and dimeric form of M2-PK (**a**) to their glycolytic substrate PEP and (**b**) to their dinucleotide substrates ADP and GDP. The tetrameric and dimeric form of M2-PK were isolated by gel permeation from MCF-7 cells

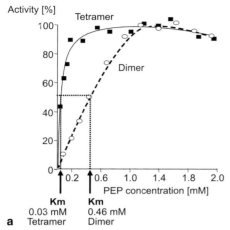
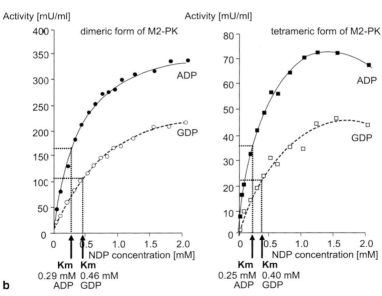

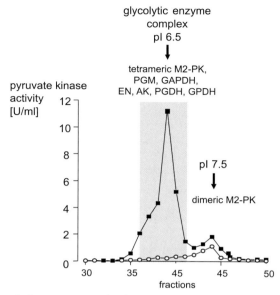

Fig. 4. Glycolytic enzyme complex in NRK cells. Migration of M2-PK into the glycolytic enzyme complex in ras-expressing NRK cells. Result of isoelectric focusing. o, parental NRK cells; ■, ras expressing NRK cells. (Mazurek et al. 2001a)

complex allows an effective conversion of glucose to pyruvate and lactate and correlates with the high rate of aerobic glycolysis described first by Otto Warburg (Warburg et al. 1924). The tumour-derived lactate lowers the pH value of the tumour environment and has been found to alter the phenotype and functional activity of dendritic cells in multicellular spheroid models (Gottfried et al. 2006).

Regarding the dinucleotide substrates, M2-PK has the highest affinity to ADP, but may also use GDP with lower affinity as a phosphate acceptor. In contrast to the PEP affinity, which is high in the case of the tetrameric form and low in the case of the dimeric form, the affinities to ADP and GDP do not differ between the tetrameric and dimeric form (Fig. 3b). The affinities to the dinucleotides UDP, CDP and TDP are low

(Mazurek et al. 1998). Accordingly, a high amount of the tetrameric form of M2-PK correlates with high ATP and GTP levels and a high ATP:ADP and GTP:GDP ratio (Fig. 1) (Zwerschke et al. 1999; Mazurek et al. 2001a,b).

The tetrameric form of M2-PK predominates in differentiated cells expressing M2-PK such as the lung. In tumour cells, M2-PK was found to be mainly in the inactive dimeric form, which at first glance appears to be inconsistent with the increased conversion of glucose to lactate described for a wide variety of tumours (Eigenbrodt and Glossmann 1980; Eigenbrodt et al. 1992). However, energy regeneration is not the only metabolic function of glycolysis in proliferating and especially tumour cells. Glycolytic intermediates are important precursors for the synthesis of cell building blocks, which are required in large amounts by proliferating cells (Fig. 1) (Eigenbrodt and Glossmann 1980; Presek et al. 1988; Eigenbrodt et al. 1992, 1998; Zwerschke et al. 1999; Mazurek et al. 2001b; Miccheli et al. 2006). Glycerate 3-P is the precursor for the synthesis of serine and glycine, C_1 units, cysteine and sphingosine. Dihydroxyacetone-P provides the backbone for phospholipids. Ribose 5-P, the sugar component of nucleotides, can be synthesized from glucose 6-P via the oxidative pentose P pathway or from fructose 6-P and glyceraldehyde 3-P via the nonoxidative pentose-P pathway, whereby studies with C14 marked glucose revealed that in tumour cells 85% of the ribose 5-P is synthesized by thiamine-dependent transketolase via the nonoxidative pentose–phosphate pathway (Boros et al. 1998). Therefore, cell proliferation is only possible if enough energy-rich phosphometabolites are available. This regulation mechanism has been termed the metabolic budget system (Eigenbrodt et al. 1992). A high level of the nearly inactive dimeric form of M2-PK as found in tumour cells induces an increase in all glycolytic phosphometabolites above the pyruvate kinase reaction and favours channelling of glucose carbons into synthetic processes. Because of its low activity, the dimeric form of M2-PK correlates with low ATP and GTP levels and low ATP:ADP and GTP:GDP ratios. On the other hand, a high level of the dimeric form correlates with high rates of nucleic acid synthesis, which is especially reflected by an increase in the UTP and CTP concentrations. Thus, cell proliferation and a high amount of the dimeric form of M2-PK was found to correlate with a low ratio between purines

(ATP+GTP) and pyrimidines (UTP+CTP), whereas a high amount of the tetrameric form of M2-PK is accompanied by a high (ATP+GTP): (UTP+CTP) ratio (Fig. 1) (Ryll and Wagner 1992; Zwerschke et al. 1999; Mazurek et al. 2001a, 2001b). When M2-PK is mainly in the inactive dimeric form and not available for glycolytic ATP production, energy can be provided by the degradation of the amino acid glutamine to glutamate, aspartate, CO_2, pyruvate, citrate and lactate, a pathway termed glutaminolysis (Lobo et al. 2000; Mazurek et al. 2001a; Rossignol et al. 2004). Glutaminolysis and the truncated citric acid cycle have the metabolic advantage that the amount of acetyl CoA infiltrated into the citric acid cycle is low and that the acetyl CoA is saved for fatty acid and cholesterol de novo synthesis. Fatty acids can be used for phospholipid synthesis or can be released. Fatty acids and glutamate are immunosuppressive and may be capable of protecting tumor cells from immune attacks (Mazurek et al. 2002).

The tetramer:dimer ratio of M2-PK is not a stationary value in tumour cells and may oscillate depending on the concentration of key metabolites as well as oncoproteins. A key regulator of the tetramer: dimer ratio of M2-PK and the metabolic budget system is the glycolytic intermediate fructose 1,6-P2 (Eigenbrodt and Glossmann 1980; Eigenbrodt et al. 1992; Ashizawa et al. 1991; Mazurek and Eigenbrodt 2003). High fructose 1,6-P2 levels induce the reassociation of the inactive dimeric form of M2-PK to the highly active tetrameric form. Consequently, glucose is converted to pyruvate and lactate with the production of energy until fructose 1,6-P2 levels drop below a critical value to allow the dissociation to the dimeric form. Another activator of M2-PK is the amino acid L-serine, which allosterically increases the affinity of M2-PK to its substrate PEP and reduces the amount of fructose 1,6-P2 necessary for tetramerization. Serine is synthesized from the glycolytic intermediate glycerate 3-P and the glutaminolytic intermediate glutamate, thereby linking both pathways (Fig. 1). Serine is an essential precursor for phospholipid and sphingolipid synthesis as well as for glycine and activated methyl groups, which are necessary substrates in purine and pyrimidine synthesis. However, if the synthesis of activated methyl groups from serine exceeds a certain rate, tetrahydrofolate is irreversibly converted to N5-methyl-tetrahydrofolate, a methyl trap, and is consequently no longer available for nucleic acid de novo synthesis.

PKM2 meets oncogenes 109

Therefore, the activation of M2-PK by serine is an effective regulatory feedback mechanism to prevent serine over-production and the methyl trap. An inhibition of M2-PK is induced by the glutaminolytic intermediate L-alanine as well as, L-cysteine, L-methionine, L-phenylalanine, L-valine, L-leucine, L-isoleucine and saturated and mono-unsaturated fatty acids. Furthermore, M2-PK is a target of the thyroid hormone 3,3',5-triiodi-L-thyronine (T_3), which binds to the monomeric form of M2-PK and prevents its association to the tetrameric form (reviewed in Eigenbrodt et al. 1992).

4 Interaction of M2-PK with Different Oncoproteins

4.1 Interaction Between M2-PK and pp60[v-src]

pp60[v-src] is a 60 kDa nonreceptor tyrosine kinase. Expression of v-src in avian and mammalian cells leads to transformation. The normal cellular homologue of v-src is the proto-oncoprotein c-src. All src kinases contain a poorly conserved unique domain at the N-terminus, three conserved Src homology domains (SH3, SH2 and SH1, whereby SH1 harbours the tyrosine kinase domain) and a C-terminal regulatory domain (Fig. 5a) (Roskoski 2004; Prakash et al. 2007). Autophosphorylation of tyrosine 419 (human c-src) within the SH1 tyrosine kinase domain is necessary for optimal activity and leads to a stabilization of the active form. Inactivation is induced by binding of phosphorylated Tyr 530 (within human C-terminal regulatory domain) to its own SH2 domain. In contrast to c-src, within the C-terminal regulating domain, the viral counterpart lacks 19 amino acids, which include the negative regulating phosphorylation site, resulting in a high level of kinase activity and a high transforming potential. V-src has been shown to phosphorylate lactate dehydrogenase, enolase and the pyruvate kinase isoenzyme type M2 both in vitro and in vivo. Whereas in vitro phosphorylation activities of pp60[v-src] and pp60[c-src] were found to be qualitatively similar, in vivo the phosphorylation activities of pp60[c-src] were only weak (Presek et al. 1980, 1988; Cooper et al. 1983; Eigenbrodt et al. 1983; Coussens et al. 1985). In chicken embryo cells, transfection with the temperature-sensitive mutant NY 68 of the Schmidt-Ruppin strain of Rous sarcoma virus induced tyrosine phosphorylation and dimerization

Fig. 5a–d. Molecular structure of oncoproteins interacting with M2-PK. **a** Human src protein: *SH*, src homology domain. **b** A-Raf protein: the conserved regions CR1 and CR2 represent the regulatory N-terminus of the enzyme. The CR3 domain harbours the catalytic activity. M2-PK binds to the very C-terminus, which is not conserved between the different Raf isoforms. **c** HPV 16-E7: deletion of aa 79–83 resulted in decreased affinity to M2-PK. *CD*, conserved domain; *CXXC*, putative zinc finger motifs. **d** HERC 1: *RLD*, RCC1 like domain; *LZ*, leucine zipper; *HECT*, Homologous to E6-AP-CArboxyl-terminus

of M2-PK within 3 h after the shift to the permissive temperature, with a maximal peak after 12 h. Similar results have been obtained with NIH 3T3 cells transfected with RSV ts LA90. The dimerization of M2-PK was accompanied by an increase in fructose 1,6-P2, P-ribose-PP and 1,2 diacylglycerol (Presek et al. 1980, 1988; Eigenbrodt et al. 1998).

4.2   Interaction Between M2-PK and A-Raf

M2-PK can also be phosphorylated in serine. In tumour cells, serine phosphorylation of M2-PK was shown to be cAMP-independent and inducible by EGF (Oude Weernink et al. 1991; Moule and McGivan 1991; Eigenbrodt et al. 1998). However, a corresponding serine kinase remained long undiscovered. The yeast two-hybrid technique revealed that M2-PK specifically interacts with the A-Raf isoenzyme. The interaction between A-Raf and M2-PK takes place within the C-terminal domain of A-Raf (Fig. 5b). The interacting region, although part of the conserved domain 3, is not conserved between the different Raf isoforms, which may explain why the two other Raf isoforms B-Raf and c-Raf were not found to interact with M2-PK within the yeast two hybrid test (Le Mellay et al. 2002). Deletion of the N-terminal regulatory domain leads to constitutive active Raf forms. A fusion between the kinase domain of A-Raf and the retroviral gag-protein (gag-A-Raf) is able to transform NIH 3T3 cells. Co-transfection of NIH 3T3 cells with a kinase dead mutant of M2-PK (M2-PK K366M) reduced colony formation of stably A-Raf-expressing NIH 3T3 cells, whereas co-transfection of NIH 3T3 cells with gag-A-Raf and wild type M2-PK led to a doubling of focus formation, which points to a cooperative effect of A-Raf

PKM2 meets oncogenes 111

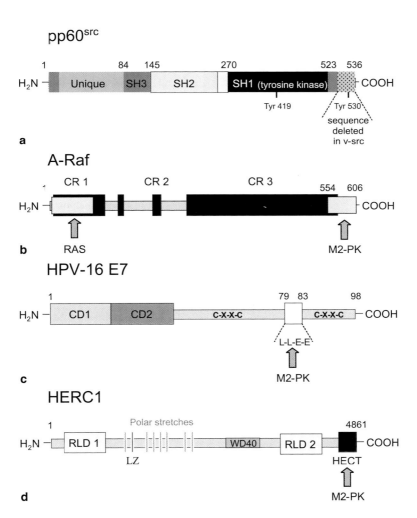

and M2-PK in cell transformation (Le Mellay et al. 2002). The effect of A-Raf on the quaternary structure of M2-PK seems to depend on the basic metabolism of the individual cell line. In primary mouse fibroblasts, which are characterized by glutamine production and serine degradation, A-Raf wild type expression induced a dimerization and inactivation of M2-PK, which resulted in a reduction of the glycolytic flux rate. In immortalized NIH 3T3 fibroblasts characterized by glutamine degradation and serine production, gag-A-Raf transformation increased the highly active tetrameric form of M2-PK and favoured degradation of glucose to lactate under the regeneration of energy. High serine levels activate M2-PK. Thus the activation and tetramerization of M2-PK found in gag-A-Raf transformed NIH 3T3 cells may be a secondary metabolic effect induced by high serine levels (Mazurek et al. 2007).

4.3 Interaction Between M2-PK and Protein Kinase C Delta

Protein kinase C delta (PKCδ) was shown to play a role in apoptosis, metastasis and tumour suppression (Kiley et al. 1999; Perletti et al. 1999; Zhong et al. 2002). It remains unknown whether PKCδ is an oncogenic or tumour suppressive protein. Two-dimensional isoelectric focusing electrophoresis in combination with MALDI mass spectroscopy identified M2-PK as a new substrate of PKCδ (Siwko and Mochly-Rosen 2007). Immunoprecipitation experiments suggest that PKCδ binds to M2-PK and rapidly releases the enzyme after phosphorylation. In vitro incubation of M2-PK with purified PKCδ neither influenced the activity nor the tetramer:dimer ratio of M2-PK. However, an in vivo effect of PKCδ on M2-PK has not yet been investigated and therefore cannot be ruled out at this point. In PKCδ$^{-/-}$ mice, an approximately twofold decrease in M2-PK levels was observed in comparison to the PKCδ$^{+/+}$, mice suggesting that phosphorylation of M2-PK by PKCδ may regulate stability or degradation of M2-PK (Mayr et al. 2004).

4.4 Interaction Between M2-PK and HPV-16 E7

The E7 oncoprotein of the human papillomavirus type 16 (HPV-16 E7) cooperates with the HPV-16 E6 oncoprotein to immortalize human keratinocytes (Münger and Howley 2002). Thus, HPV-16 belongs to the high-risk types of human papillomavirus and is linked to malignant human cervix cancer (zur Hausen 2002). HPV-16 E7 consists of 98 amino acids and contains two conserved domains CD1 and CD2 at the N-terminus and two zinc finger motifs (C-X-X-C) at the carboxy terminus (Fig. 5c). The conserved domain 2 (CD2) within the N-terminus mediates binding of E7 to proteins of the retinoblastoma gene family, thereby contributing to the deregulation of the cell cycle (Münger and Howley 2002). The carboxy terminus of HPV-16 E7 acts as an interaction domain for M2-PK. Deletion of the amino acids 79–83 (Leu, Leu, Glu, Glu) within the HPV-16 E7 protein leads to a reduced affinity of HPV-16 E7 to M2-PK and reduces the transforming potential of E7, suggesting that binding of M2-PK may play a role in cell transformation (Zwerschke et al. 1999). Thus, the transforming activity of E7 is sensitive to mutations in both the N-terminus as well as the C-Terminus (Jewers et al. 1992). NIH 3T3 cells which are already immortal are transformed by E7 alone, whereas transformation of primary normal rat kidney cells (NRK) require expression of a second oncoprotein ras (Zwerschke et al. 1999; Mazurek et al. 2001a,b). The parental NRK cells were characterized by low glycolytic enzyme activities and a low glycolytic flux rate. The stable expression of ras induced an increase in most of the glycolytic enzymes, including 6-phosphofructo 1 kinase (PFK) and M2-PK as well as an increase in the glycolytic flux rate. The increase in PFK activity correlated with an increase of fructose 1,6-P2 levels, which resulted in a tetramerization and migration of M2-PK into the glycolytic enzyme complex in close proximity to adenylate kinase (AK) (Fig. 4). The close association between the highly active M2-PK and AK led to a decrease in ATP and an increase in AMP levels. High AMP levels inhibit cell proliferation by inhibiting P-ribose PP synthetase, a key enzyme in purine and pyrimidine synthesis (Mazurek et al. 1997). Accordingly, in ras-expressing cells, cell proliferation was inhibited. However, the expression of ras dramatically boosts tumour metabolism, thereby preparing the metabolome of the cells for transformation. In E7-transformed cells, binding of E7 to M2-PK induced a dimerization and migration of M2-PK out of the glycolytic enzyme complex which favoured the channelling of glucose carbons into synthetic processes. Accordingly, UTP and CTP levels increased, whereas ATP and GTP levels decreased in E7 transformed cells.

4.5 Interaction Between M2-PK and HERC1

HERC 1, also termed oncH according to its identification in a nude mouse tumorigenicity assay and p532 according to its molecular weight, is one of four proteins within the human HERC protein family (Rosa et al. 1996). HERC proteins contain a HECT (homologous to E 6 AP carboxyl terminus) domain in their carboxyl-terminus and one or more RCC1-like domains (RLDs) elsewhere in their amino acid sequence. RCC1 (regulator of chromosome condensation 1-protein) is a guanine nucleotide exchange factor (GEF) for RAN, a small GTP-binding protein which is predominantly located in the nucleus and involved in the nuclear transport of proteins with nuclear localization signals. HERC 1, which was shown to be consistently over-expressed in several tumour cell lines, contains two RLD domains (Fig. 5d). RLD1 is a GEF for ARF-1, Rab3a and Rab5, which are all three GTPases involved in cellular membrane trafficking (Rosa et al. 1996). RLD2, for which as yet no GEF activity has been shown, specifically binds to ARF-1 in the Golgi apparatus as well as to clathrin and Hsp70 (Rosa and Barbacid 1997). The yeast two hybrid technique, in vitro pull-down experiments, as well as in vivo pull-down experiments in Sf9 insect cells infected with baculovirus encoding full-length M2-PK and the His-tagged HECT domain of HERC1 (last 366 aa of HERC1), revealed that M2-PK specifically binds to the HECT domain of the HERC1 protein (Garcia-Gonzalo et al. 2003). The M2-PK sequence involved in HERC1 binding contains the critical residues for fructose 1,6-P2 binding as well as for the intersubunit contact. HECT domains confer E3 ubiquitin protein ligase activity and are involved in protein degradation. However, all results so far appear to indicate that the interaction of M2-PK with HERC1 influences neither M2-PK activity nor the tetramer:dimer ratio of M2-PK, nor does it induce ubiquitination and increased degradation of M2-PK (Garcia-Gonzalo et al. 2003). Therefore, the physiological function of the interaction between M2-PK and HERC1 is still not known. Since M2-PK also phosphorylates GDP, it is conceivable that M2-PK may function as a local GTP producer (nano machine) for the RLDs as well as for the GTPases ARF-1 and Rab5.

5   Role of M2-PK in the Nucleus

M2-PK contains an inducible nuclear translocation signal (NLS) in its C-domain, which, in contrast to classical NLS, is not rich in arginine and lysine (Hoshino et al. 2007). The role of M2-PK within the nucleus is complex since pro-proliferative as well as pro-apoptotic stimuli have been described. In BB13 cells, an interleukin 3-dependent haematopoietic cell line, which ectopically expresses the EGF receptor, IL-3 stimulation induced a translocation of M2-PK into the nucleus within 30 min. The IL-3 stimulated nuclear translocation of M2-PK was dependent on JAK2. In the same cell system, the over-expression of a construct of the M2-PK protein fused with the NLS from SV40-T antigen enhanced EGF-stimulated cell proliferation in the absence of IL-3 (Hoshino et al. 2007). The mechanism by which nuclear M2-PK enhances cell proliferation is yet not clear. In Morris hepatoma 7777 tumour cells, nuclear M2-PK was found to participate in the phosphorylation of histone 1 by direct phosphate transfer from PEP to histone 1 (Ignacak and Stachurska 2003). On the other hand, nuclear translocation of M2-PK induced by the somatostatin analogue TT 232, H_2O_2 or UV light has recently been linked to the induction of a caspase-independent programmed cell death (Stetak et al. 2007).

6   Tumour M2-PK: A Biomarker for Metabolic Profiling of Tumours

Measurements of v-max activities in different cell lines allowed the classification of proliferating cell lines in the following three groups: nontumour normal proliferating cells with PK-v-max activities between 30 and 950 mU/mg protein, tumour cell lines with PK-v-max activities between 900 and 1300 mU/mg protein and metastatic tumour cell lines with PK-v-max activities between 1590 and 1630 mU/mg protein (Board et al. 1990). V-max activities of enzymes are measured at saturated substrate concentrations. In the case of M2-PK v-max, activities were measured at saturated PEP concentrations, which means that both the tetrameric and dimeric form are highly active. In contrast, at physiological PEP concentrations, only the tetrameric form of M2-PK

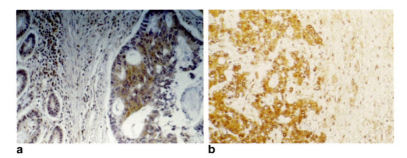

Fig. 6a,b. Immunohistology staining of (**a**) rectum carcinoma and (**b**) metastases of the rectum carcinoma in the liver with the monoclonal anti M2-PK antibody clone DF4. (Hardt et al. 2004a; Mazurek 2008)

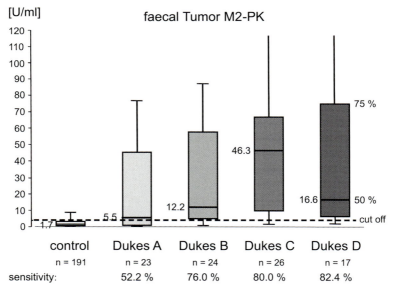

Fig. 7. Correlation between faecal tumour M2-PK and TNM staging. (Hardt et al. 2004b)

PKM2 meets oncogenes

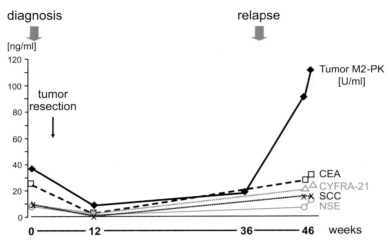

Fig. 8. Follow-up study of a patient with a squamous carcinoma ($T_2N_1M_0$) of the lung. After surgical resection, a local relapse surrounding the arteria pulmonalis and metastases in infracranial and paratracheal lymph nodes occurred between weeks 36 and 46. (Schneider et al. 2002)

is highly active, whereas the dimeric form is nearly inactive (Fig. 3a). Immunohistological staining of various tumours with monoclonal antibodies which specifically recognize the dimeric form of M2-PK allows the visualization of the pyruvate kinase isoenzyme shift in tumour cells. This technique shows that the distribution of M2-PK in primary tumours can be heterogeneous, whereas their metastases are always stained very homogeneously (Fig. 6).

Furthermore, the dimeric form of M2-PK (tumour M2-PK) is released from tumour cells into the blood and from gastrointestinal tumours also into the stool of tumour patients, most likely by tumour necrosis and cell turnover, providing the possibility of diagnostic application. Thus, the amount of tumour M2-PK was found to increase in the EDTA plasma of patients with renal cell carcinoma, melanoma, lung, breast, cervical, ovarian, oesopharyngeal, gastric, pancreatic and colorectal cancer as well as in stool samples of patients with gastric and colorectal cancer and to correlate with tumour stages (Fig. 7) (Wechsel et al. 1999; Lüftner et al. 2000; Schneider et al. 2002; Hardt et al. 2004b; Kaura et al. 2004; Ahmed et al. 2007; Koss et al. 2008; Kumar et al. 2007).

Therefore, tumour M2-PK is an organ-unspecific biomarker which reflects the metabolic activity and proliferation capacity of tumours. An important field of application of the plasma test are follow-up studies to monitor failure, relapse or success during therapy (Fig. 8).

Interestingly, in different human gastric carcinoma cell lines, cisplatin resistance was found to correlate with low M2-PK protein levels and activity (Yoo et al. 2004). Low PK activities promote synthetic processes debranching from glycolysis, such as the oxidative pentose P-shuttle, an important source for NADPH production within cells. NADPH is necessary for reduction of glutathione (GSSG) and activation of the thioredoxin system, both of which have been shown to be involved in cisplatin resistance.

7   Conclusions

The expression of the pyruvate kinase isoenzyme type M2, which can switch between a highly active tetrameric form and a nearly inactive dimeric form, is an important metabolic sensor to adapt tumour metabolism to different metabolic conditions, such as nutrient supply. The quantification of the dimeric form of M2-PK in plasma and stool is a tool for early detection of tumours and therapy control.

Acknowledgements. This chapter is dedicated to Prof. Dr. Erich Eigenbrodt, head of the Comparative Biochemistry of Animals Department within the Veterinary Faculty of the University of Giessen, who significantly contributed to our knowledge of the role of M2-PK within the tumour metabolome and diagnosis and passed away in 2004.

References

Ahmed AS, Dew T, Lawton FG, Papadopoulos AJ, Devaja O, Raju KS, Sherwood RA (2007) M2-PK as a novel marker in ovarian cancer: a prospective cohort study. Eur J Gynaec Oncol 28:83–88

PKM2 meets oncogenes

Ashizawa K, Willingham MC, Liang CM, Cheng SY (1991) In vivo regulation of monomer-tetramer conversion of pyruvate kinase subtype M2 by glucose is mediated via fructose 1,6-bisphosphate. J Biol Chem 266:16842–16846

Board M, Humm S, Newsholme EA (1990) Maximum activities of key enzymes of glycolysis, glutaminolysis, pentose phosphate pathway and tricarboxylic acid cycle in normal, neoplastic and suppressed cells. Biochem J 265:503–509

Boros LG, Lee PW, Brandes JL, Cascante M, Muscarella P, Schirmer WJ, Melvin WS, Ellison EC (1998) Nonoxidative pentose phosphate pathways and their direct role in ribose synthesis in tumors: is cancer a disease of cellular glucose metabolism? Med Hypothesis 50:55–59

Brahimi-Horn MC, Pouyssegur J (2007) Oxygen a source of life and stress. FEBS Lett 581:3582–3591

Brinck U, Eigenbrodt E, Oehmke M, Mazurek S, Fischer G (1994) L- and M2-pyruvate kinase expression in renal cell carcinomas and their metastases. Virchows Arch 424:177–185

Brugge JS, Erikson RL (1977) Identification of a transformation-specific antigen induced by an avian sarcoma virus. Nature 269:346–348

Cooper JA, Reiss NA, Schwartz RJ, Hunter T (1983) Three glycolytic enzymes are phosphorylated at tyrosine in cells transformed by Rous sarcoma virus. Nature 302:218–223

Coussens PM, Cooper JA, Hunter T, Shalloway D (1985) Restriction of the in vitro and in vivo tyrosine protein kinase activities of pp60c-src relative to pp60v-src. Mol Cell Biol 5:2753–2763

Discher DJ, Bishopric NH, Wu X, Peterson CA, Webster KA (1998) Hypoxia regulates β-enolase and pyruvate kinase-M promoters by modulation Sp1/Sp3 binding to a conserved GC element. J Biol Chem 273:26087–26093

Dombrauckas JD, Santarsiero BD, Mesecar AD (2005) Structural basis for tumor pyruvate kinase M2 allosteric regulation and catalysis. Biochemistry 44:9417–9429

Eigenbrodt E, Glossmann H (1980) Glycolysis – one of the keys to cancer? Trends Pharmacol Sci 1:240–245

Eigenbrodt E, Fister P, Rübsamen H, Friis RR (1983) Influence of transformation by Rous sarcoma virus on the amount, phosphorylation and enzyme kinetic properties of enolase. EMBO J 2:1565–1570

Eigenbrodt R, Reinacher M, Scheefers-Borchel U, Scheefers H, Friis RR (1992) Double role of pyruvate kinase type M2 in the expansion of phosphometabolite pools found in tumor cells. In: Perucho M (ed) Critical reviews in oncogenesis. CRC Press, Boca Raton, FL, pp. 91–115

Eigenbrodt E, Mazurek S, Friis R (1998) Double role of pyruvate kinase type M2 in the regulation of phosphometabolite pools. In: Bannasch P, Kanduc D, Papa S, Tager JM (eds) Cell growth and oncogenesis. Birkhäuser Verlag, Basel, pp. 15–30

Garcia-Gonzalo FR, Cruz C, Munoz P, Mazurek S, Eigenbrodt E, Ventura F, Bartrons R, Rosa JL (2003) Interaction between HERC1 and M2-type pyruvate kinase. FEBS Lett 539:78–84

Gatenby RA, Gillies RJ (2004) Why do cancers have high aerobic glycolysis? Nat Rev Cancer 4:891–899

Gottfried E, Kunz-Schughart LA, Ebner S, Müller-Klieser W, Hoves S, Andreesen R, Mackensen A, Kreutz M (2006) Tumor-derived lactic acid modulates dendritic cell activation and antigen expression. Blood 107:2013–2021

Hardt PD, Mazurek S, Klör HU, Eigenbrodt E (2004a) Neuer Test zum Nachweis von Darmkrebs. Spiegel der Forschung 21:15–19

Hardt PD, Mazurek S, Toepler M, Schlierbach P, Bretzel RG, Eigenbrodt E, Kloer HU (2004b) Faecal tumour M2 pyruvate kinase: a new, sensitive screening tool for colorectal cancer. Br J Cancer 91:980–984

Hentze MW (1994) Enzymes as RNA-binding proteins: a role for (di)-nucleotide-binding domains? Trends Biochem Sci 19:101–103

Hoshino A, Hirst JA, Fujii H (2007) Regulation of cell proliferation by interleukin-3-induced nuclear translocation of pyruvate kinase. J Biol Chem 282:17706–17711

Hunter T, Sefton BM (1980) Transforming gene product of Rous sarcoma virus phosphorylates tyrosine. Proc Natl Acad Sci U S A 77:1311–1315

Ignacak J, Stachurska MB (2003) The dual activity of pyruvate kinase type M2 from chromatin extracts of neoplastic cells. Comp Biochem Physiol Part B 134:425–433

Jewers RJ, Hildebrandt P, Ludlow JW, Kell B, McCance DJ (1992) Regions of human papillomavirus type 16 E7 oncoprotein required for immortalization of human keratinocytes. J Virol 66:1329–1335

Kaura B, Bagga R, Patel FD (2004) Evaluation of the pyruvate kinase isoenzyme tumor (Tu M2-PK) as a tumor marker for cervical carcinoma. J Obstet Gynaecol Res 30:193–196

Kiley SC, Clark KJ, Goodnough M, Welch DR, Jaken S (1999) Protein kinase C delta involvement in mammary tumor cell metastasis. Cancer Res 59:3230–3238

Koss K, Maxton D, Jankowski JA (2008) Faecal dimeric M2 pyruvate kinase in colorectal cancer and polyps correlates with tumour staging and surgical intervention. Colorectal Dis 10:244–248

Kress S, Stein A, Maurer P, Weber B, Reichert J, Buchmann A, Huppert P, Schwarz M (1998) Expression of hypoxia-inducible genes in tumor cells. J Cancer Res Clin Oncol 124:315–320

Kumar Y, Tapuria N, Kirmani N, Davidson BR (2007) Tumour M2-pyruvate kinase: a gastrointestinal cancer marker. Eur J Gastroenterol Hepatol 19:265–276

Le Mellay V, Houben R, Troppmair J, Hagemann C, Mazurek S, Frey U, Beigel J, Weber C, Benz R, Eigenbrodt E, Rapp UR (2002) Regulation of glycolysis by A-Raf protein serine/threonine kinase. Adv Enzyme Regul 42:317–332

Lobo C, Ruiz-Bellido MA, Aledo JC, Marquez J, Nunez de Castro I, Alonso FJ (2000) Inhibition of glutaminase expression by antisense mRNA decreases growth and tumourigenicity of tumor cells. Biochem J 348:257–261

Lüftner D, Mesterharm J, Akrivakis C, Geppert R, Petrides PE, Wernecke KD, Possinger K (2000) Tumor M2-pyruvate kinase expression in advanced breast cancer. Anticancer Res 20:5077–5082

Mayr M, Chung YL, Mayr U, McGregor E, Troy H, Bayer G, Leitges M, Dunn MJ, Griffiths JR, Xu Q (2004) Loss of PKC-delta alters cardiac metabolism. Am J Pysiol Heart Circ Physiol 287:H937–H945

Mazurek S (2008) Das Tumor-Metabolom – eine Quelle von Messgrößen zur frühzeitigen Diagnose von Tumoren. In: Hardt PD (ed) Tumormarker in der Gastroenterologie. Unimed Verlag, Bremen, pp 55–65

Mazurek S, Eigenbrodt E (2003) The tumor metabolome. Anticancer Res 23:1149–1154

Mazurek S, Michel A, Eigenbrodt E (1997) Effect of extracellular AMP on cell proliferation and metabolism of breast cancer cell lines with high and low glycolytic rates. J Biol Chem 272:4941–4952

Mazurek S, Grimm H, Wilker S, Leib S, Eigenbrodt E (1998) Metabolic characteristics of different malignant cancer cell lines. Anticancer Res 18:3275–3282

Mazurek S, Zwerschke W, Jansen-Dürr P, Eigenbrodt E (2001a) Effects of the human papilloma virus HPV-16 E7 oncoprotein on glycolysis and glutaminolysis: role of pyruvate kinase and the glycolytic enzyme complex. Biochem J 356:247–256

Mazurek S, Zwerschke W, Jansen-Dürr P, Eigenbrodt E (2001b) Metabolic cooperation between different oncogenes during cell transformation: interaction between activated ras and HPV-16 E7. Oncogene 20:6891–6898

Mazurek S, Grimm H, Boschek CB, Vaupel P, Eigenbrodt E (2002) Pyruvate kinase type M2: a crossroad in the tumor metabolome. Br J Nutr 87:S23–S29

Mazurek S, Drexler H, Troppmair J, Eigenbrodt E, Rapp UR (2007) Regulation of pyruvate kinase type M2 by A-Raf: a possible stop or go mechanism. Anticancer Res 27:3963–3971

Miccheli A, Tomassini A, Puccetti C, Valerio M, Peluso G, Tuccillo F, Calvani M, Manetti C, Conti F (2006) Metabolic profiling by 13C-NMR spectroscopy: [1,2–13C2]glucose reveals a heterogeneous metabolism in human leukemia T cells. Biochimie 88:437–448

Moule SK, McGivan JD (1991) Epidermal growth factor stimulates the phosphorylation of pyruvate kinase in freshly isolated rat hepatocytes. FEBS Lett 280:37–40

Münger K, Howley PM (2002) Human papillomavirus immortalization and transformation functions. Virus Res 89:213–228

Nagy E, Rigby WF (1995) Glyceraldehyde 3-P dehydrogenase selectively binds AU-rich RNA in the NAD^+-binding region (Rossmann Fold). J Biol Chem 270:2755–2763

Noguchi T, Inoue H, Tanaka T (1986) The M1 and M2-type isoenzymes of rat pyruvate kinase are produced from the same gene by alternative RNA splicing. J Biol Chem 261:13807–13812

Noguchi T, Yamada K, Inoue H, Matsuda T, Tanaka T (1987) The L- and R-type isozymes of rat pyruvate kinase are produced from a single gene by use of different promotors. J Biol Chem 262:14366–14371

Oude Weernink PA, Rijksen G, Staal GEJ (1991) Phosphorylation of pyruvate kinase and glycolytic metabolism in three human glioma cell lines. Tumor Biol 12:339–352

Perletti GP, Marras E, Concari P, Piccinini F, Tashjian AH (1999) PKCdelta acts as growth and tumor suppressor in rat colonic epithelial cells. Oncogene 18:1251–1256

Prakash O, Bardot SF, Cole JT (2007) Chicken sarcoma to human cancers: a lesson in molecular therapeutics. Ochsner J 7:61–64

Presek P, Glossmann H, Eigenbrodt E, Schoner W, Rübsamen H, Friis RR, Bauer H (1980) Similarities between a phosphoprotein (pp60src)-associated protein kinase of Rous sarcoma virus and a cyclic adenosine 3′:5′-monophosphate independent protein kinase that phosphorylates pyruvate kinase type M2. Cancer Res 40:1733–1741

Presek P, Reinacher M, Eigenbrodt E (1988) Pyruvate kinase type M2 is phosphorylated in tyrosine residues in cells transformed by Rous sarcoma virus. FEBS Lett 242:194–198

Reinacher M, Eigenbrodt E (1981) Immunohistological demonstration of the same type of pyruvate kinase isoenzyme (M2-PK) in tumors of chicken and rat. Virchows Arch B Cell Pathol Incl Mol Pathol 37:79–88

Rosa JL, Barbacid M (1997) A giant protein that stimulates guanine nucleotide exchange on ARF1 and Rab proteins forms a cytosolic ternary complex with clathrin and Hsp70. Oncogene 15:1–6

Rosa JL, Casaroli-Marano RP, Buckler AJ, Vilaro S, Barbacid M (1996) p619, a giant protein related to the chromosome condensation regulator RCC1, stimulates guanine nucleotide exchange on ARF1 and Rab proteins. EMBO J 15:4262–4273; Corrigendum 1996: EMBO J 15:5738

Roskoski R (2004) Src protein-tyrosine structure and regulation. Biochem Biophys Res Commun 324:1155–1164

Rossignol R, Gilkerson R, Aggeler R, Yamagata K, Remington SJ, Capaldi RA (2004) Energy substrate modulates mitochondrial structure and oxidative capacity in cancer cells. Cancer Res 64:985–993

Rous P (1910) A transmissible avian neoplasm. (Sarcoma of the common Fowl). J Exp Med 12:696–705

Ryll T, Wagner R (1992) Intracellular ribonucleotide pools as a tool for monitoring the physiological state of in vitro cultivated mammalian cells during production processes. Biotechnol Bioeng 40:934–946

Schneider J, Neu K, Grimm H, Velcovsky HG, Weisse G, Eigenbrodt E (2002) Tumor M2-pyruvate kinase in lung cancer patients: immunohistochemical detection and disease monitoring. Anticancer Res 22:311–318

Shim H, Dolde C, Lewis BC, Wu CS, Dang G, Jungmann RA, Dalla-Favera R, Dang CV (1997) c-Myc transactivation of LDH A: implications for tumor metabolism and growth. Proc Natl Acad Sci U S A 94:6658–6663

Siwko S, Mochly-Rosen D (2007) Use of a novel method to find substrates of protein kinase C delta identifies M2 pyruvate kinase. Int J Biochem Cell Biol 39:978–987

Staal GEJ, Rijksen G (1991) Pyruvate kinase in selected human tumors. In: Pretlow TG, Pretlow TP (eds) Biochemical and molecular aspects of selected cancers. Academic Press, San Diego, pp 313–337

Steinberg P, Klingelhöffer A, Schäfer A, Wüst G, Weisse G, Oesch F, Eigenbrodt E (1999) Expression of pyruvate kinase M2 in preneoplastic hepatic foci of N-nitrosomorpholine-treated rats. Virchows Arch 434:213–220

Stetak A, Veress R, Ovadi J, Csermely P, Keri G, Ullrich A (2007) Nuclear translocation of the tumor marker pyruvate kinase M2 induces programmed cell death. Cancer Res 67:1602–1608

Stubbs M, Bashford CL, Griffiths JR (2003) Understanding the tumor metabolic phenotype in the genomic era. Curr Mol Med 3:49–59

Warburg O, Poesener K, Negelein E (1924) Über den Stoffwechsel der Karzinomzellen. Biochem Z 152:309–344

Wechsel HW, Petri E, Bichler KH, Feil G (1999) Marker for renal carcinoma (RCC): the dimeric form of pyruvate kinase type M2 (Tu M2-PK). Anticancer Res 19:2583–2590

Yamada K, Noguchi T (1999) Regulation of pyruvate kinase M gene expression. Biochem Biophys Res Commun 256:257–262

Yoo BC, Ku JL, Hong SH, Shin YK, Park SY, Kim HK, Park JG (2004) Decreased pyruvate kinase M2 activity linked to cisplatin resistance in human gastric carcinoma cell lines. Int J Cancer 108:532–539

Zhong M, Lu Z, Foster DA (2002) Downregulating PKC delta provides a PI3K/Akt-independent survival signal that overcomes apoptotic signals generated by c-src overexpession. Oncogene 21:1071–1078

Zur Hausen H (2002) Papillomaviruses and cancer: from basic studies to clinical applications. Nat Rev Cancer 2:342–350

Zwerschke W, Mazurek S, Massimi P, Banks L, Eigenbrodt E (1999) Modulation of type M2 pyruvate kinase activity by the human papillomavirus type 16 E7 oncoprotein. Proc Natl Acad Sci U S A 96:1291–1296

Ernst Schering Foundation Symposium Proceedings, Vol. 4, pp. 125–152
DOI 10.1007/2789_2008_092
© Springer-Verlag Berlin Heidelberg
Published Online: 16 July 2008

Molecular Imaging of Tumor Metabolism and Apoptosis

U. Haberkorn[✉], A. Altmann, W. Mier, M. Eisenhut

Department of Nuclear Medicine, University of Heidelberg, INF 400, 69120 Heidelberg, Germany
email: *uwe.haberkorn@med.uni-heidelberg.de*

1	Glucose Metabolism	126
2	Amino Acids	132
3	Apoptosis	140
4	Hypoxia	141
References		144

Abstract. Increased metabolism has been found to be one of the most prominent features of malignant tumors. This property led to the development of tracers for the assessment of glucose metabolism and amino acid transport and their application for tumor diagnosis and staging. Prominent examples are fluorodeoxyglucose, methionine and tyrosine analogs, which have found broad clinical application. Since quantitative procedures are available, these techniques can also be used for therapy monitoring. Another approach may be based on the noninvasive detection of apoptosis with tracers for phosphatidyl-serine presentation and/or caspase activation as surrogate markers for therapeutic efficacy. Finally, the evaluation of hypoxia with nitroimidazoles may be a valuable tool for prognosis and therapy planning.

1   Glucose Metabolism

Malignant tumors are tissues metabolizing glucose to lactate to a high extent. This increased glycolytic activity correlates with a high amount of mitochondrial-bound hexokinase in the tumor cells. In quickly growing tumor cells, the hexokinase activity is greatly enhanced, and up to 80% of the molecules are bound to the outer mitochondrial membrane. Changes in the expression of glycolysis-associated genes during the malignant transformation have been reported by several groups (Shawver et al. 1987; Flier et al. 1987; Bos et al. 2002): especially the gene encoding the glucose transporter subtype 1 (GLUT1) is activated early after transformation of cells with oncogenes such as src, ras or fps. An increase in the mRNA of GLUT1 is observed as early as 4–6 h after induction of the p21 c-H-ras oncoprotein, while morphological changes occur after 72–76 h. Furthermore, the increase in GLUT1 mRNA after ras transfection was independent of the growth rate. In vivo overexpression of GLUT1 and GLUT3 was found in a series of different human and experimental tumors. The increase in GLUT1 transcription can be used for imaging or therapy by cloning a reporter gene or a therapeutic gene such as suicide genes downstream of the GLUT1 promoter/enhancer elements (Haberkorn et al. 2002, 2005). Examples are the herpes simplex virus thymidine kinase (HSVtk) gene or the sodium iodide symporter, where adeno associated virus or retroviral vectors have been used to transfect tumor cells and to measure the uptake of specific substrates or to treat animals with genetically modified tumors (Sieger et al. 2003, 2004). In these studies, reporter gene expression (green fluorescent protein, HSVtk or sodium iodide symporter) was specific for tumor cells or cells with expression of an activated ras oncogene (Fig. 1).

[18]Fluordeoxyglucose (FDG) for PET studies of glucose metabolism was introduced as a consequence of autoradiographic and biochemical studies with glucose analogs in different tissues. Similar to glucose, 2-deoxyglucose (dGlc) and FDG are transported bidirectionally and are phosphorylated by the enzyme hexokinase. This is possible because the C-2 position, unlike the C-1, C-3 and C-6 positions, is uncritical for the binding to the hexokinase. In contrast to glucose-6-phosphate, FDG-6-phosphate and dGlc-6-phosphate are not further metabolized in significant amounts during the examination. dGlc-6-phosphate is not

Molecular Imaging of Tumor Metabolism and Apoptosis

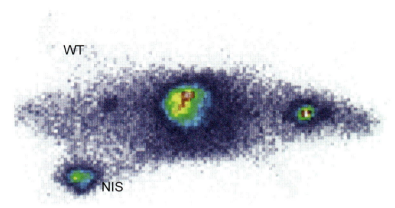

Fig. 1. Scintigraphic image of a rat bearing a wild type (*WT*) and a genetically modified tumor with expression of the human sodium iodide symporter (*NIS*). Only the NIS-expressing tumor, the stomach and the thyroid gland show accumulation of ^{131}I

metabolized to fructose-6-phosphate and, therefore, is not a substrate for the glucose-6-phosphate dehydrogenase. dGlc-6-phosphate may be converted to dGlc-1-phosphate and uridine-diphosphate(UDP)-dGlc, followed by an incorporation into glycogen, glycolipids and glycoproteins. However, these reactions are very slow in mammalian tissues. Furthermore, in the brain, the organ where the deoxyglucose method was applied for the first time, as well as in malignant tumors, glucose-6-phosphatase activity is downregulated. In contrast to the autoinhibition of the glucose phosphorylation, FDG-6-phosphate shows no inhibition of hexokinase activity. Compared to 2-deoxyglucose, FDG is incorporated very slowly into macromolecules, as has been demonstrated in yeasts as well as in chick fibroblasts. Due to their negative charge, which prevents penetration of the negatively charged inner part of the plasma membrane, FDG-6-phosphate and dGlc-6-phosphate accumulate in the cells. A further advantage is the rapid clearance of the tracer: similar to glucose, FDG shows glomerular filtration. However, unlike glucose, this is not followed by tubular reabsorption, because FDG is not a substrate for the tubular sodium glucose symporter, which is responsible for the rapid renal clearance.

PET studies with different animal models showed a correlation of FDG uptake and the content of GLUT1 and hexokinase mRNA (Haberkorn et al. 1994). Differences in the FDG uptake in different lung carcinomas, with lower values for adenocarcinomas as compared to squamous cell carcinomas, corresponded to the histologically determined expression of GLUT1, which was higher in squamous cell carcinomas than in adenocarcinomas. Therefore, the genetic program in malignant tumors leads to the corresponding FDG uptake values as measured with PET. Similar results were obtained in bronchioalveolar adenocarcinomas, with significantly lower values for the number of GLUT1-positive cells and FDG uptake and a correlation of histologic grade and the amount of GLUT1-positive cells and FDG uptake.

The clinical application of [18]FDG was predominantly for tumor diagnosis and staging for a variety of tumor entities such as lung, colon, breast, head and neck, and esophageal cancer, melanoma, and lymphoma (Fig. 2). In lung cancer, a meta-analytic comparison of PET (14 studies, 514 patients) and CT (29 studies, 2226 patients) for the demonstration of mediastinal nodal metastases in patients with non-small cell lung cancer (NLCLC) was done by Dwamena et al. (1999). In this analysis, pooled point estimates of diagnostic performance and summary ROC curves indicated that PET was significantly more accurate than CT for demonstration of nodal metastases with $p < 0.001$. The mean sensitivity and specificity were 0.79 ± 0.03 and 0.91 ± 0.02, respectively, for PET and 0.60 ± 0.02 and 0.77 ± 0.02, respectively, for CT. Subgroup analyses did not alter these findings. The results were collected and evaluated in a consensus conference leading to recommended applications of the method for a variety of tumors (Reske and Kotzerke 2001; Tables 1 and 2).

Besides staging, the prognostic value of FDG-PET has also been evaluated. The relation of high pretherapeutic FDG uptake to a poorer

Fig. 2. Transaxial PET/CT images of a patient with lung cancer. *Top:* CT image showing a large hilar mass. *Middle:* The fusion image demonstrates the smaller extent of the tumor and atelectatic lung tissue. *Bottom:* A metastasis in the adrenal gland is visualized

Molecular Imaging of Tumor Metabolism and Apoptosis

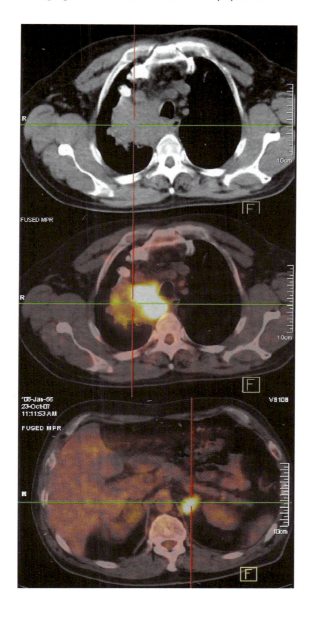

Table 1 Indications of FDG-PET with established or probable clinical value

Differentiated thyroid carcinoma	Restaging of radioiodine-negative lesions
	Restaging of radioiodine-positive lesions
Esophageal carcinoma	Staging lymph nodes, distant metastases
Pancreatic cancer	DD inflammation/tumor
	Recurrence
Colorectal carcinoma	Therapy monitoring
	Recurrence
Mammary carcinoma	N-staging
Head and neck tumors	N-staging
	Recurrence
	CUP
Lung tumors	Solitary pulmonary nodule
	N-staging (NSCLC)
	Extrathoracic N-staging
	Recurrence
Hodgkin lymphoma	Staging
	Therapy monitoring
Highly malignant NHL	Staging
	Therapy monitoring
Melanoma	N-staging (Breslow > 1.5 mm or known lymph node metastases)
	M-staging (Breslow > 1.5 mm or known lymph node metastases)
	Recurrence
	Follow-up for pT3 and pT4 tumors
	Follow-up of metastases
Bone/soft tissue tumors	Dignity, biological behavior, surgery planning

Table 2 Indications for a clinical value of FDG-PET in single cases

Mammary carcinoma	Dignity
	M-staging
Ovarian carcinoma	Recurrence
Head and neck tumors	Detection of a second tumor
Lung tumors	Therapy monitoring

Molecular Imaging of Tumor Metabolism and Apoptosis

prognosis was observed by different groups in patients with lung cancer (Ahuja et al. 1998; Vansteenkiste et al. 1999). In a study with 155 patients, the uptake in the primary lesion of NSCLC was compared to the clinical outcome: independent of other clinical findings, patients with higher uptake values had a shorter median survival time than patients with lower FDG accumulation (Ahuja et al. 1998). In this respect, a correlation was described between tumor growth and FDG uptake (Duhaylongsod et al. 1995). However, in experimental studies, conflicting reports exist concerning the possible correlation of FDG uptake and tumor cell proliferation (Brown et al. 1999; Higashi et al. 2000).

PET using ^{18}F-FDG has been applied for the evaluation of treatment response during chemotherapy, gene therapy, and radiotherapy in a variety of tumors, indicating that FDG delivers useful parameters for the early assessment of therapeutic efficacy (Bassa et al. 1996; Haberkorn et al. 1991, 1993, 1997a,b, 1998; Rozenthal et al. 1989). Furthermore, tumors may react to therapeutic intervention by compensatory reactions, including an increase in glucose metabolism, especially during the very early phase after treatment.

In general, increased FDG transport rates early after treatment are suggested as evidence of stress reactions in tumors after chemotherapy, gene therapy or radiation therapy (Haberkorn et al. 1998, 2001). The glucose carrier shows a complex regulation: glucose transport may be altered by phosphorylation of the transport protein (Hayes et al. 1993), decreased degradation (Shawver et al. 1987), translocation from intracellular pools to the plasma membrane (Widnell et al. 1990), or an increased expression of the gene (Flier et al. 1987). The increase in glucose transport after exposure of cells to damaging agents has been ascribed mainly to a redistribution of the glucose transport protein from intracellular pools to the plasma membrane. Such reactions have been found in cells exposed to arsenite, calcium ionophore A23187; or 2-mercaptoethanol (Widnell et al. 1990; Wertheimer et al. 1991; Hughes et al. 1989). Furthermore, increased glucose metabolism has been observed after chemotherapy or gene therapy of hepatoma with HSV thymidine kinase (Haberkorn et al. 1998, 2001a,b). Incubation with cytochalasin B or deoxyglucose after the end of treatment increased the amount of apoptotic cells (Haberkorn et al. 2001a,b), whereas monotherapy with these drugs had no effect. Enhanced glycolysis may be used for a metabolic design of combination therapy, as has been done for chemotherapy (Haberkorn et al. 1992) or radiotherapy (Singh et al. 2005). These strategies intend to disturb possible repair processes that are in need of energy by interfering with glycolysis. Besides deoxyglucose, a few compounds are available such as 6-aminonicotinamide, 3-bromopyruvate, oxythiamine, 5-thioglucose, or genistein, where at least deoxyglucose shows a rather selective toxicity for cells with chemotherapy resistance (Haberkorn et al. 1992). The design of such a combination treatment requires data on the changes in the metabolic pathways with respect to dose and time dependence, which may be obtained by FDG-PET.

2 Amino Acids

Although PET with ^{18}F-fluorodeoxyglucose (FDG) has been proven to be useful for diagnosis and therapy monitoring in a variety of tumors, there is a need for complementary information of tumor biology. FDG is not tumor-selective and shows accumulation in inflammatory lesions. Furthermore, tissues with high background such as the brain may cause difficulties in image interpretation. Malignant tumors show changes in the amino acid transport, protein synthesis and proliferation. Therefore, many efforts have been made to establish tracers based on amino acids or proliferation markers.

Radiolabeled amino acids are used for measuring the rate of protein synthesis and amino acid transport. Besides protein synthesis, amino acids are precursors for many other biomolecules, such as adenine, cytosine, histamine, thyroxine, epinephrine, melanin and serotonin, and are important in other metabolic cycles, including transamination and transmethylation; methionine has a specific role in the initiation of protein synthesis and amino acids such as glutamine are used for energy. Since all these pathways create a dependency on amino acid uptake, amino acid transport does not faithfully represent protein synthesis, but rather provides a general measure of the cellular need for amino acids.

Amino acids enter cells mainly via specific transport systems (Christensen 1990). These systems can be sodium-dependent or -independent. Sodium-dependent transport relies on the sodium chemical gradient and

Molecular Imaging of Tumor Metabolism and Apoptosis

the electric potential across the plasma membrane, as well as on the activity of the Na^+/K^+-ATPase. Sodium-independent systems depend on the amino acid concentration gradient across the cell membrane and are often coupled to the countertransport (i.e., in the opposite direction) of K^+.

Kinetic studies have identified several sodium-dependent transport systems: A, ASC and Gly, which transport amino acids with short polar or linear side chains, e.g., alanine, serine and glycine. In general, a change in affinity occurs when a sodium ion binds to the transporter protein. Subsequent binding of the amino acid results in a conformational change in the transporter protein and in turn to the influx of the attached sodium ion and the amino acid into the cell. System A is transinhibited by intracellular substrates (i.e., the presence of intracellular substrates slows the uptake of amino acids), whereas system ASC is trans-stimulated by the presence of intracellular substrates (i.e., the presence of intracellular substrates increases the activity).

Sodium-independent systems, L (ubiquitously found), $B^{0,+}$ and y^+, are carriers for branched chain and aromatic amino acids, e.g., leucine, valine, tyrosine and phenylalanine. System L shows trans-stimulation by intracellular substrates such as leucine and valine. Most amino acid carrier systems can also transport synthetic, nonmetabolizable amino acid analogs.

Regulation of amino acid transport is complex and is influenced by hormones, cytokines, changes in cell volume and the availability of nutrients (Christensen 1990). For example, the number of system A active carriers increases during starvation; hence patients should be studied preferentially while fasting.

Malignant cells were found to have an increased amino acid transport (Boerner et al. 1985; Busch et al. 1959; Isselbacher 1972; Saier et al. 1988). Strong expression of system A has been found in transformed and malignant cells as a result of oncogene action (Saier et al. 1988) and a correlation between amino acid transport and cellular proliferation has been described (Jager et al. 2000; Kuwert et al. 1997).

For the assessment of the protein synthesis rate, relatively complex kinetic models are necessary. Furthermore, there is no existing constant correlation between the quantitative data derived from these models and the grade of malignancy (Ogawa et al. 1993). Although ^{11}C-leucine appears to be the best amino acid for measuring protein synthesis rate (Vaalburg et al. 1992), most studies have used methionine because of the ease of tracer synthesis. The drawbacks of methionine are its use in metabolic cycles other than protein synthesis, which results in a variety of metabolites and difficulties in quantification (Ishiwata et al. 1989, 1996). Conflicting reports have been published about the specificity of carrier-mediated transport of methionine into brain tumors in studies comparing D- and L-methionine using an overload of branched amino acids. Furthermore, at least part of the tracer uptake seems to be the result of passive diffusion. Cellular uptake in vitro is mainly accomplished via the L system with minor contributions from systems A and ASC.

Patient studies have revealed high uptake of methionine in the pituitary gland and pancreas, moderate uptake in salivary glands, lacrimal glands and bone marrow, and low uptake in the normal brain (Jager et al. 2001). It has been used as a tracer mainly in brain tumors, where it shows excellent contrast between normal brain and tumors and high sensitivity for tumor detection (Herholz et al. 1998; Langen et al. 1997). Also, in a study of 196 patients, the accuracy of differentiation between low- and high-grade lesions was 79% (Herholz et al. 1998). Tumor delineation was better than with CT, MRI and FDG-PET (Mosskin et al. 1986; Bergstrom et al. 1983; Kaschten et al. 1998).

A high sensitivity for the detection of primary and metastatic brain tumors was also found using either ^{11}C-tyrosine or L-2-^{18}F-fluorotyrosine (Wienhard et al. 1991; Willemsen et al. 1995). Analysis of the plasma metabolites of ^{11}C-tyrosine revealed that ^{11}C-CO_2, ^{11}C-proteins and ^{11}C-L-DOPA constituted more than 50% of total plasma radioactivity at 40 min after injection making a complex pharmacokinetic model for further analysis necessary. Using a five-compartment model, it was shown that while the net protein synthesis rate was dependent on the recycling of amino acids from protein, tracer influx into the cell was not. The curve-fitting results of dynamic scans were unreliable because of the exchange of ^{11}C-tyrosine between plasma and erythrocytes, whereas the graphical Patlak-Gjedde analysis was not influenced by this. L-2-^{18}F-fluorotyrosine was studied in 15 patients with brain tumors and showed rapid uptake, which was mainly attributed to an increase in transport. Also, an improved localization of tumor tissue for biopsy has been described for both methionine and tyrosine.

Molecular Imaging of Tumor Metabolism and Apoptosis

In head and neck cancer, amino acids have been used mainly for staging. Primary tumors have shown higher methionine uptake as compared to surrounding tissues, and tumors larger than 1 cm in diameter have been detected with a sensitivity of 91% (Leskinen-Kallio et al. 1994). Noninvasive tumor grading has not been possible (Lindholm et al. 1998). Tyrosine-PET has shown comparable results with a significantly higher protein synthesis rate for tumor as compared to nontumor tissue (deBoer et al. 2002).

Lung cancer has also shown high uptake of methionine with high sensitivity, but low specificity, for solitary pulmonary nodules (Kubota et al. 1990). Staging was improved by methionine in a retrospective study, but gave no advantage as compared to FDG-PET.

Comparisons of FDG and amino acids in patients with breast cancer have revealed that FDG was better than tyrosine, but not as good as methionine (Jansson et al. 1995). In lymphoma, no association between methionine uptake with histologic grade has been seen, unless kinetic analysis was applied (Rodriguez et al. 1995). Differentiation between benign and malignant lesions has also been possible for soft tissue sarcomas using tyrosine-PET (Plaat et al. 1999). Studies with small patient numbers have been conducted either with methionine or tyrosine in patients with melanoma, bladder cancer, metastatic nonseminoma, ovarian cancer and uterine cancer.

Since ^{123}I-α-methyl tyrosine (^{123}I-IMT) has proven to be a promising SPECT tracer for imaging amino acid transport in tumors (Fig. 3), ^{124}I-IMT and L-3-^{18}F-α-methyl tyrosine (FMT) have been synthesized for PET studies (Amano et al. 1998; Langen et al. 1990). ^{124}I-IMT accumulates in brain and tumor tissue, reaching a maximum concentration after 15 min with a washout of 20%–35% at 60 min after injection. Animal experiments have confirmed the accumulation of the intact tracer in brain without incorporation of the tracer into proteins. FMT uptake was high in the pancreas and in several tumor models. Tumor uptake of FMT was reduced by inhibition of the amino acid transport systems. The tumor:blood ratios of FMT in mice with LS180 (human colon cancer), RPMI1788 (human B-cell lymphoma) and MCF7 (human mammary carcinoma) tumors at 60 min after injection were 1.8, 5.9 and 3.6, respectively. Most of the activity was localized in the acid-soluble fraction, suggesting that FMT is mainly not incorporated into proteins.

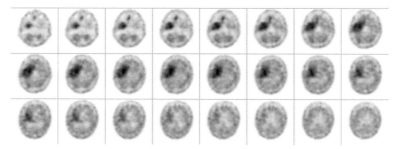

Fig. 3. Transaxial PET/CT images of a patient with lung cancer. *Top:* CT image showing a large hilar mass. *Middle:* The fusion image demonstrates the smaller extent of the tumor and atelectatic lung tissue. *Bottom:* A metastasis in the adrenal gland is visualized

Clinical studies have shown that brain tumors are better delineated by FMT as compared with FDG, with no dependence of FMT uptake on tumor grade (Inoue et al. 1999). In contrast, FMT uptake correlated with histologic grade in musculoskeletal tumors, but with a better discriminative capacity for FDG.

O-(2-^{18}F-fluoroethyl)- L-tyrosine (L-FET), which is also not incorporated into proteins, has been evaluated in mammary carcinoma-bearing mice and in mice with the colon carcinoma cell line SW707 (Wester et al. 1999; Heiss et al. 1999). Results of transport inhibition experiments with specific competitive inhibitors have demonstrated that the uptake of L-FET into SW707 cells is caused mainly by system L. In vivo studies revealed a plasma half-life of L-FET of 94 min and increasing brain uptake up to 120 min with a brain:blood ratio of 0.9. Xenotransplanted tumors have shown higher uptake of L-FET (> 6% injected dose/g) than all other organs, except the pancreas. High-performance liquid chromatography (HPLC) analysis of brain, pancreas and tumor homogenates as well as plasma samples of mice at 10, 40 or 60 min after injection revealed only unchanged L-FET, indicating high stability and lack of metabolization of the tracer. Preliminary clinical results are available for high-grade brain tumors and metastatic melanomas (Weber et al. 2001).

Molecular Imaging of Tumor Metabolism and Apoptosis

Once transported into the cell, tyrosine is metabolized to dihydroxy-phenylalanine (DOPA), which can be used for melanin synthesis. Therefore, DOPA labeled with ^{18}F at the 2 position has been used for tumor characterization in melanoma-bearing mice (Ishiwata et al. 1991). Tumors with a lower melanin synthesis rate accumulated less DOPA than tumors with a higher rate. The metabolite predominantly found in these studies was ^{18}F-MeFDOPA. DOPA labeled at the 6 position is commonly used for the evaluation of the dopaminergic system. It has also been used to study patients with metastatic melanoma; tracer uptake was perfusion-independent with DOPA-PET showing a lower sensitivity as compared to FDG-PET (Dimitrakopoulou-Strauss et al. 2001).

A variety of synthetic amino acids, including α-aminoisobutyric acid (AIB), 1-aminocyclopentane carboxylic (ACPC) acid, 2-amino-3-fluoro-2-methylpropanoic acid (FAMP), 3-fluoro-2-methyl-2-(methyl-amino)propanoic acid (N-MeFAMP) and 1-amino-3-fluorocyclobutane-1-carboxylic acid (FACBC), have been synthesized and evaluated, mostly in cell culture and animal systems. AIB is thought to be actively accumulated in viable cells primarily by the A-type amino acid transport system and has shown avid uptake in a melanoma model. Additionally, ACPC and AIB imaging were found to be superior to FDG in C6 gliomas and Walker 256 rat carcinosarcoma, especially for identifying tumor infiltration of adjacent brain tissue beyond the macroscopic border of the tumor, and in low-grade tumors with an intact blood–brain barrier. Contrast-enhancing regions of the tumors were visualized more clearly with AIB than with FDG or Ga-DTPA; viable and necrotic-appearing tumor regions could be distinguished more readily with AIB than with FDG (Uehara et al. 1997). Increased AIB uptake was also observed in soft tissue sarcomas (Schwarzbach et al. 1999). As for AIB, amino acid transport assays using 9L gliosarcoma cells demonstrated that FAMP and N-MeFAMP are substrates for the A type amino acid transport system and show very high tumor:normal brain ratios: 36:1 and 104:1, respectively (McConathy et al. 2002). In a rat brain tumor model, maximum tumor uptake of ^{18}F-FACBC was seen at 60 min, with a tumor:normal brain ratio of 5.6 at 5 min and 6.6 at 60 min after tracer administration (Shoup et al. 1999).

Measurement of the effects of therapy on tumor metabolism may be useful in predicting therapy outcome at an early stage of treatment. This principle may be applied not only to glucose metabolism but also to amino acid transport and metabolism. Studies of different human tumors treated with a variety of therapies and of the rat AH109A tumor model after radiotherapy demonstrated a rapid posttherapeutic reduction in methionine uptake, reflecting inactivation of protein synthesis and damage to the membrane transport system (Jansson et al. 1995; Bergstrom et al. 1987; Schaider et al. 1996). Furthermore, the uptake of L-1-^{11}C-tyrosine in rhabdomyosarcoma of Wag/Rij rats was dose-dependently reduced after local hyperthermia (Daemen et al. 1991). Moreover, the accumulation of AIB is decreased in rat prostate tumors after long-term treatment with stilbestrol (Dunzendorfer et al. 1981). These changes were followed later by a reduction in tumor mass.

In vitro studies have demonstrated that methotrexate and cisplatin induce a *decline* in AIB and methionine accumulation in L1210 murine leukemia cells (Scanlon et al. 1983, 1987), leading to the speculation that the inhibition of methionine uptake by methotrexate may be due to drug binding to a specific membrane carrier, or a reduction in the sodium gradient across the plasma membrane, which is necessary for the uptake of amino acids, or effects on intracellular processes which support uptake of amino acids. Higashi et al. demonstrated an *increase* in methionine and FDG uptake in human ovarian carcinoma cells after radiotherapy, which was accompanied by an increase in cell volume (Higashi et al. 1993). These phenomena were interpreted as giant cell formation with enlarged cellular volume and continued protein synthesis, but accelerated repair was also suggested. Another in vitro study combined the information obtained from experiments using a transport tracer (AIB) and a tracer that is transported and metabolized (methionine) and found a decrease in neutral amino acid transport after gene therapy of hepatoma cells with HSV thymidine kinase and ganciclovir, indicating treatment effects on the energy-dependent transport systems (Haberkorn et al. 1997a). Methionine uptake experiments showed a decrease in tracer accumulation in the acid-insoluble fraction (representing nucleic acids and proteins), indicating impaired protein synthesis and an increase in the acid-soluble fraction. The increase in radioactivity in the acid-soluble fraction may be caused by enhanced transmethy-

Molecular Imaging of Tumor Metabolism and Apoptosis

lation processes, which usually are observed during oncogenic transformation and after exposure to DNA-damaging agents.

Clinical studies in brain tumors have been done for early evaluation of treatment response and differentiation between recurrence and radiation necrosis. In ten patients with low-grade gliomas, a dose-dependent reduction in methionine uptake was seen after brachytherapy (Wurker et al. 1996). Differentiation between radiation necrosis and tumor recurrence was possible with methionine-PET (Ogawa et al. 1991). In another study with ten patients, tyrosine-PET showed no change in the protein synthesis rate despite a decrease in tumor volume in seven patients (Heesters et al. 1998).

After radiotherapy of head and neck cancer, a lower posttherapeutic methionine uptake was shown to correlate with therapy response (Lindholm et al. 1998). Similar results were obtained in patients after radiotherapy or chemotherapy of lung, breast and rectal cancer (Jansson et al. 1995; Daemen et al. 1991). However, the predictive value of methionine-PET remains questionable.

Amino acids have been suggested to be useful in the differentiation between inflammation and malignancy. Experimental studies have shown that amino acids accumulate less than FDG in inflammation (Kubota et al. 1989). However, uptake may occur in benign lesions such as ischemic brain, infarction, scar, abscesses and sarcoidosis, and also in irradiated areas. Therefore, active inflammatory cells may need amino acids and the specificity of amino acids for tumor imaging is not absolute. However, in mice with tumor-infiltrated or inflammatory lymph nodes, the accumulation of O-(2-^{18}F-fluoroethyl)-L-tyrosine showed significant differences with no overlap between inflammatory and tumorous nodes (Rau et al. 2002).

In summary, amino acids may have a potential role in the characterization of the biological properties of tumors as increased amino acid transport or protein synthesis. Advantages over FDG imaging can be expected in the imaging of brain tumors, because the background of tracer accumulation is lower than FDG. The role of amino acids for the monitoring of tumor response to treatment as well as the differentiation between inflammation and tumor tissue has to be established in further studies.

3   Apoptosis

For the in vivo detection of apoptosis, two main targets in the apoptotic pathway are of interest: (1) the presentation of phosphatidylserine residues at the outer side of the plasma membrane and (2) the appearance of activated caspases (Martin et al. 1995; Villa et al. 1997). Phosphatidylserine is maintained at the inner site of the plasma membrane by the adenosine triphosphate (ATP)-dependent enzymes floppase and translocase (Zwaal and Schroit 1997). Apoptosis induced inactivation of these enzymes and activation of a scramblase leads to the appearance of phosphatidylserine on the outer side of the membrane. This effect has been recently used to develop an imaging agent for apoptosis (Blankenberg et al 1998, 1999): Annexin V, a 35-kDa human protein with high affinity for cell membrane-bound phosphatidylserine, was labeled with ^{99m}Tc and investigated for its uptake in apoptotic cells. An increased accumulation was found in Jurkat cells where the programmed cell death was initiated by growth factor deprivation, anti-CD95 antibody and doxorubicin treatment. Also, anti-CD95 treated mice showed a threefold rise in hepatic ^{99m}Tc-Annexin V accumulation in response to severe liver damage with histologic evidence of apoptosis. Finally, increased uptake was detected in animal models using the acute rejection of transplanted heterotopic cardiac allografts or transplanted murine B cell lymphomas treated with cyclophosphamide (Blankenberg et al. 1999).

Since caspases play a key role during the early period of the intracellular signal cascade of cells undergoing apoptosis, benzyloxycarbonyl-Val-Ala-DL-Asp(O-methyl)-fluoromethyl ketone [Z-VAD-fmk], a pan-caspase inhibitor, was evaluated as a potential apoptosis imaging agent (Haberkorn et al. 2001c). Uptake measurements were made with Morris hepatoma cells (MH3924Atk8), which showed expression of the herpes simplex virus thymidine kinase (HSVtk) gene. Apoptosis was induced by treatment of the cells with ganciclovir and a twofold increase of $[^{131}I]$I-Z-VAD-fmk uptake was found at the end of treatment with the HSVtk/suicide system, which consistently remained elevated for the following 4 h. The slow cellular influx and lack of uptake saturation of $[^{131}I]$IZ-VAD-fmk are evidence for simple diffusion as a transport mechanism. In addition, the absolute cellular uptake of $[^{131}I]$IZ-VAD-fmk was found to be low. Instead of using an inhibitor, synthetic

Molecular Imaging of Tumor Metabolism and Apoptosis

caspase substrates that may accumulate in the apoptotic cell by metabolic trapping, thereby enhancing the imaging signal are currently being investigated. In a recent study, ten radiolabeled peptides containing the DEVDG sequence, selective for downstream caspases such as caspase-3, were synthesized and evaluated for their uptake kinetics using an apoptosis test system (Bauer et al. 2005). Within this series of peptides, radioiodinated Tat49–57-yDEVDG-NH2 and Tat57–49-yDEVDG-NH2, both containing an additional HIV Tat sequence, were taken up by apoptotic cells to a significantly higher extent than with the controls. The enhanced uptake was interpreted as the interaction of the labeled peptide or fragment with activated caspases. Current efforts are focused on alternative radioisotopes that include radiometal complexes to further improve these characteristics.

4 Hypoxia

Because of uncontrolled growth and a misbalance between tumor mass and vascularization, oxygen limitation is a common feature of malignant tumors. Oxygen concentration inside solid tumors is reduced, which contributes to the tumor aggressiveness and poor prognosis of patients (Stadler et al. 1999). Genomes of tumor cells become unstable under hypoxic conditions, and hypoxia can be the selective pressure for the expansion of clones with anti-apoptotic treatment-resistant or highly metastatic potential (Young et al. 1988; Graeber et al. 1996). Resistance to chemotherapy and radiation therapy can be attributed, at least in part, to the hypoxic condition of tumor cells (Teicher 2004). Hypoxia confers these aggressive properties on the tumors through either the remodeling of tumor vasculature or the direct phenotypic changes of tumor cells themselves. Tumor cells under hypoxia can acquire anti-apoptotic and chemoresistant properties through changes in the expression of apoptosis-related molecules. Furthermore, the involvement of HIF-1α in the tumor progression to an anti-apoptotic phenotype was reported (Erler et al. 2004).

Oxygen deprivation is encountered by the induction of various genes. Hypoxia inducible factor 1 (HIF) plays a central role in this regulatory system. HIF can induce the production of a variety of gene products relevant for metabolism, vascularization, survival, pH and cell migration. Active HIF-1 is a heterodimer composed of two subunits, HIF-1α and HIF-1β. HIF-1β is constitutively expressed independent of environmental oxygen concentration, while the expression of HIF-1α is negligible under normoxia and induced under hypoxia. Up to now, HIF-1α, HIF-2α and HIF-3α have been identified and cloned as the members of HIF α family that can dimerize with HIF-1α and bind to hypoxia responsible elements (HRE) in the genes of hypoxia-responsive molecules.

Among HIFα family members, HIF-1α is thought to be the key molecule regulating the cellular response to physiological and pathological hypoxia. Mechanisms of hypoxia-induced expression of HIF-1α have been intensively studied, and the intracellular level of HIF-1α protein under reduced oxygen concentration was found to be increased mainly through stabilization of the protein. Turnover of the HIF-1α protein is regulated by the ubiquitin–proteasome system, in which target proteins are degraded by proteasome depending on their ubiquitylation (Semenza 2002). Under normoxia, the level of the HIF-1α protein is kept low through rapid ubiquitylation and subsequent proteasomal degradation. HIF-1α becomes susceptible to rapid ubiquitylation through hydroxylation of proline residues at Pro-402 and Pro-564 by prolyl hydroxylase 2 (PHD2), which requires oxygen for its enzyme activity (Berra et al. 2003). In cells under hypoxia, the ubiquitylation and subsequent degradation of HIF-1α is suppressed because of the decrease in PHD2 activity, and therefore the level of HIF-1a protein increases. In addition, the activity of HIF-1 as a transcription factor is also controlled by hydroxylation of HIF-1α protein. Hydroxylation of asparagine residue at Asn-803 inhibits the interaction between HIF-1α and p300, which is essential for the transcriptional activity of HIF-1 (Lando et al. 2002b). Because the factor inhibiting HIF (FIH) that hydroxylates Asn-803 is also an oxygen-dependent enzyme, the transcriptional activity of HIF-1 increases under hypoxia due to the suppressed hydroxylation at Asn-803 (Lando et al. 2002a; Hewitson et al. 2002). Cells can control the transcription of HIF-1-regulated genes by sensing the oxygen concentration through the activities of oxygen-dependent enzymes PHD2 and FIH, and consequently regulating the intracellular level as well as the transcriptional activity of HIF-1 (Haddad 2002).

Molecular Imaging of Tumor Metabolism and Apoptosis

Although HIF-1 can be activated by nonhypoxic pathways, hypoxia inside the growing tumor mass is the most probable candidate for the activation of HIF-1α cascade in tumor cells. This is supported by the fact that both HIF-1α and VEGF expression are upregulated predominantly in tumor cells around the necrotic areas of highly vascularized tumor mass in glioblastoma (Plate et al. 1992). Therefore, angiogenesis triggered by the hypoxia-HIF-1α-VEGF cascade seems to play an important role in tumor progression to the more aggressive phenotypes.

The noninvasive imaging of hypoxic areas may be used for the development of individualized therapies, new therapeutic approaches or as a prognostic marker. At present, oxygen partial pressure (pO_2) is measured with the Eppendorf probe, which showed significant correlations of pO_2 and therapy response in clinical studies. This method has several limitations: its application is restricted to lesions located at surface areas and its invasiveness precludes it from being done routinely or repeatedly. Furthermore, differentiating between areas of necrosis and areas with anoxic/hypoxic but living cells is not possible.

Nitroimidazoles are reduced under hypoxic conditions by intracellular reductases to reactive intermediate metabolites. This process is dependent on the hypoxia level and may lead up to a 40-fold increase in the amount of reduced products. The metabolites bind covalently to thiol moieties of intracellular proteins, leading to an accumulation in living hypoxic cells. The resulting complexes can then be detected with antibodies, MRS, flow cytometry, autoradiography and scintigraphy or PET. 2-nitroimidazole can be labeled with ^{18}F, ^{123}I, [131]I (iodinated azomycinarabinoside, IAZA) and [99m]Tc. In vitro studies and animal experiments showed the selectivity of [^{18}F]Fluoromisonidazole for hypoxic cells. Tracer accumulation was quantitated with mathematical models as well as by determining the SUV. The results obtained so far show that [^{18}F]FMISO uptake measurements underestimate the amount of hypoxia at very low pO_2-values (2–3 mmHg). This is probably caused by the fact that below a defined cutoff level the reduction processes can no longer be increased.

Clinically, [^{18}F]FMISO was applied for the assessment of myocardial ischemia, tumor hypoxia in head and neck tumors, gliomas, non-small cell lung tumors and in soft tissue sarcomas (Padhani et al. 2007; Lee and Scott 2007). Quantitation was done by determining the hypoxic fraction volume (FHV), which is defined as the procentual fraction of tumor pixels showing a tracer accumulation at 2 h after infection at least 1.4-fold higher than the activity in plasma. After radiation therapy, a significant decrease of the FHV has been observed. However, the tracer uptake was not dependent on the tumor size, grading or VEGF expression. In patients with non-small cell lung cancer or head and neck cancer, 97% of the tumors showed accumulation of the tracer, with a great variability in the extent in different tumor entities, however, but also in different lesions from the same patient.

4-[^{18}F]Fluoro-2,3-dihydroxy-1–2(2′-nitro-1′-imidazolyl)butane([^{18}F] Fluoroerythroimidazol, [^{18}F]FETNIM) showed higher tumor:blood and tumor:muscle ratios in animal experiments than [^{18}F]FMISO (Grönros et al. 2004). The tracer accumulates strongly in liver and tumor, with no binding to plasma proteins and no peripheral metabolization. In patients with head and neck tumors, better tumor:muscle ratios were obtained in comparison to [^{18}F]FMISO.

Preliminary results have been reported for the evaluation of ^{62}Cu-labeled diacetyl-bis(N4-methylthiosemicarbazone) (^{62}Cu-ATSM) as a possible hypoxia imaging agent (Padhani et al. 2007). ^{62}Cu-ATSM showed a rapid clearance from the blood in all patients, with a low uptake in lung tissue and an intense accumulation in tumors. Furthermore, a negative correlation was found between blood flow and the flow-normalized ^{62}Cu-ATSM uptake in three out of four patients. This was interpreted as evidence for an increased ^{62}Cu-ATSM accumulation under conditions of low blood flow.

In summary, all these imaging procedures may be used to characterize the biological features of tumors and their metastases with respect to metabolism, apoptosis and microenvironment. The information obtained with these techniques can be expected to individualize treatment and make radioisotope-based methods promising tools for tumor detection, therapy planning, and therapy monitoring.

References

Ahuja V, Coleman RE, Herndon J, Patz EF (1998) Prognostic significance of FDG-PET imaging in patients with non-small cell lung cancer. Cancer 83:918–924

Molecular Imaging of Tumor Metabolism and Apoptosis

Amano S, Inoue T, Tomiyoshi K et al (1998) In vivo comparison of PET and SPECT radiopharmaceuticals in detecting breast cancer. J Nucl Med 39:1424–1427

Bassa P, Kim EE, Inoue T, Wong FC, Korkmaz M, Yang DJ, Hicks KW, Buzdar AU, Podoloff DA (1996) Evaluation of preoperative chemotherapy using PET with fluorine-18-fluorodeoxyglucose in breast cancer. J Nucl Med 37:931–938

Bauer C, Bauder-Wuest U, Mier W, Haberkorn U, Eisenhut M (2005) ^{131}I-labeled peptides as caspase substrates for apoptosis imaging. J Nucl Med 46:1066–1074

Bergstrom M, Collins VP, Ehrin E et al (1983) Discrepancies in brain tumor extent as shown by computed tomography and positron emission tomography using [^{68}Ga]EDTA, [^{11}C]glucose, and [^{11}C]methionine. J Comput Assist Tomogr 7:1062–1066

Bergstrom M, Muhr C, Lundberg PO et al (1987) Rapid decrease in amino acid metabolism in prolactin-secreting pituitary adenomas after bromocriptine treatment: a PET study. J Comput Assist Tomogr 11:815–819

Berra E, Benizri E, Ginouves A, Volmat V, Roux D, Pouyssegur J (2003) HIF prolyl-hydroxylase 2 is the key oxygen sensor setting low steady-state levels of HIF-1alpha in normoxia. EMBO J 22:4082–4090

Blankenberg FG, Katsikis PD, Tait JF, Davis RE, Naumovski L, Ohtsuki K et al (1998) In vivo detection and imaging of phosphatidylserine expression during programmed cell death. Proc Natl Acad Sci U S A 95:6349–6354

Blankenberg FG, Katsikis PD, Tait JF, Davis RE, Naumovski L, Ohtsuki K et al (1999) Imaging of apoptosis (programmed cell death) with 99mTc annexin V. J Nucl Med 40:184–191

Boerner P, Saier MH (1985) Adaptive regulatory control of system A transport activity in a kidney epithelial cell line (MDCK) and in a transformed variant. J Cell Physiol 122:308–315

Bos R, van Der Hoeven JJ, van Der Wall E, van der Groep P, van Diest PJ, Comans EFI, Joshi U, Semenza GL, Hoekstra OS, Lammertsma AA, Molthoff CFM (2002) Biologic correlates of (18)fluorodeoxyglucose uptake in human breast cancer measured by positron emission tomography. J Clin Oncol 20:379–387

Brown RS, Leung JY, Kison PV, Zasadny KR, Flint A, Wahl RL (1999) Glucose transporters and FDG uptake in untreated primary human non-small cell lung cancer. J Nucl Med 40:556–565

Busch H, Davis JR, Honig GR et al (1959) The uptake of a variety of amino acids into nuclear proteins of tumors and other tissues. Cancer Res 19:1030–1039

Christensen HN (1990) Role of amino acid transport and countertransport in nutrition and metabolism. Physiol Rev 70:43–76

Daemen BJ, Elsinga PH, Mooibroek J et al (1991) PET measurements of hyperthermia-induced suppression of protein synthesis in tumors in relation to effects on tumor growth. J Nucl Med 32:1587–1592

DeBoer JR, vander Laan BFAM, Oruim J et al (2002) Carbon-11 tyrosine PET for visualization and protein synthesis rate assessment of laryngeal and hypopharyngeal carcinomas. Eur J Nucl Med 29:1182–1187

Dimitrakopoulou-Strauss A, Strauss LG, Burger C (2001) Quantitative PET studies in pretreated melanoma patients: a comparison of 6-[^{18}F]fluoro-L-DOPA with ^{18}F-FDG and ^{15}O-water using compartment and noncompartment analysis. J Nucl Med 42:248–256

Duhaylongsod FG, Lowe VJ, Patz EF Jr, Vaughn AL, Coleman RE, Wolfe WG (1995) Lung tumor growth correlates with glucose metabolism measured by fluoride-18 fluorodeoxyglucose positron emission tomography. Ann Thorac Surg 60:1348–1352

Dunzendorfer U, Schmall B, Bigler RE et al (1981) Synthesis and body distribution of alpha-aminoisobutyric acid-L-^{11}C in normal and prostate cancer bearing rat after chemotherapy. Eur J Nucl Med 6:535–538

Dwamena BA, Sonnad SS, Angobaldo JO, Wahl RL (1999) Metastases from non-small cell lung cancer: mediastinal staging in the 1990s – meta-analytic comparison of PET and CT. Radiology 213:530–536

Erler JT, Cawthorne CJ, Williams KJ et al (2004) Hypoxia-mediated downregulation of Bid and Bax in tumors occurs via hypoxia inducible factor 1-dependent and -independent mechanisms and contributes to drug resistance. Mol Cell Biol 24:2875–2889

Flier JS, Mueckler MM, Usher P, Lodish HF (1987) Elevated levels of glucose transport and transporter messenger RNA are induced by ras or src oncogenes. Science 235:1492–1495

Graeber TG, Osmanian C, Jacks T et al (1996) Hypoxia-mediated selection of cells with diminished apoptotic potential in solid tumors. Nature 379:88–91

Grönroos T, Bentzen L, Marjamäki P, Murata R, Horsman MR, Keiding S, Eskola O, Haaparanta M, Minn H, Solin O (2004) Comparison of the biodistribution of two hypoxia markers [^{18}F]FETNIM and [^{18}F]FMISO in an experimental mammary carcinoma. Eur J Nucl Med Mol Imaging 31:513–520

Haberkorn U, Eisenhut M (2005) Molecular imaging and therapy – a programme based on the development of new biomolecules. Eur J Nucl Med Mol Imaging 32:1354–1359

Molecular Imaging of Tumor Metabolism and Apoptosis

Haberkorn U, Strauss LG, Dimitrakopoulou A, Engenhart R, Oberdorfer F, Ostertag H, Romahn J, van Kaick G (1991) PET studies of fluorodeoxyglucose metabolism in patients with recurrent colorectal tumors receiving radiotherapy. J Nucl Med 32:1485–1490

Haberkorn U, Reinhardt M, Strauss LG, Oberdorfer F, Berger MR, Altmann A, Wallich R, Dimitrakopoulou A, van Kaick G (1992) Metabolic design of combination therapy: use of enhanced fluorodeoxyglucose uptake caused by chemotherapy. J Nucl Med 33:1981–1987

Haberkorn U, Strauss LG, Dimitrakopoulou A et al (1993) Fluorodeoxyglucose imaging of advanced head and neck cancer after chemotherapy. J Nucl Med 34:12–17

Haberkorn U, Ziegler SI, Oberdorfer F, Trojan H et al (1994) FDG uptake, tumor proliferation and expression of glycolysis associated genes in animal tumor models. Nucl Med Biol 21:827–834

Haberkorn U, Altmann A, Morr I et al (1997a) Multi tracer studies during gene therapy of hepatoma cells with HSV thymidine kinase and ganciclovir. J Nucl Med 38:1048–1054

Haberkorn U, Bellemann ME, Altmann A, Gerlach L, Morr I, Oberdorfer F, Brix G, Doll J, Blatter J, Kaick G van (1997b) PET 2-fluoro-2-deoxyglucose uptake in rat prostate adenocarcinoma during chemotherapy with gemcitabine. J Nucl Med 38:1215–1221

Haberkorn U, Bellemann ME, Gerlach L, Morr I, Trojan H, Brix G, Altmann A, Doll J, van Kaick G (1998) Uncoupling of 2-fluoro-2-deoxyglucose transport and phosphorylation in rat hepatoma during gene therapy with HSV thymidine kinase. Gene Ther 5:880–887

Haberkorn U, Altmann A, Kamencic H, Morr I, Traut U, Henze M, Jiang S, Metz J, Kinscherf R (2001a) Glucose transport and apoptosis after gene therapy with HSV thymidine kinase. Eur J Nucl Med 28:1690–1696

Haberkorn U, Bellemann ME, Brix G, Kamencic H, Morr I, Traut U, Altmann A, Doll J, Blatter J, Kinscherf R (2001b) Apoptosis and changes in glucose transport early after treatment of Morris hepatoma with gemcitabine. Eur J Nucl Med 28:418–425

Haberkorn U, Kinscherf R, Krammer PH, Mier W, Eisenhut M (2001c) Investigation of a potential scintigraphic marker of apoptosis: radioiodinated Z-Val-Ala-DL-Asp(O-Methyl)-fluoromethyl ketone. Nucl Med Biol 28:793–798

Haberkorn U, Altmann A, Eisenhut M (2002) Functional genomics and proteomics – the role of nuclear medicine. Eur J Nuc Med 29:115–132

Haberkorn U, Mier W, Eisenhut M (2005) Scintigraphic imaging of gene expression and gene transfer. Curr Med Chem 12:779–794

Haddad JJ (2002) Oxygen-sensing mechanisms and the regulation of redox-responsive transcription factors in development and pathophysiology. Respir Res 3:26

Hayes N, Biswas C, Strout HV, Berger J (1993) Activation by protein synthesis inhibitors of glucose transport into L6 muscle cells. Biochem Biophys Res Commun 190:881–887

Heesters MA, Go KG, Kamman RL et al (1998) ^{11}C-tyrosine positron emission tomography and ^{1}H magnetic resonance spectroscopy of the response of brain gliomas to radiotherapy. Neuroradiology 40:103–108

Heiss P, Mayer S, Herz M et al (1999) Investigation of transport mechanism and uptake kinetics of O-(2-[^{18}F]fluoroethyl)-L-tyrosine in vitro and in vivo. J Nucl Med 40:1367–1373

Herholz K, Holzer T, Bauer B et al (1998) ^{11}C-methionine PET for differential diagnosis of low-grade gliomas. Neurology 50:1316–1322

Hewitson KS, McNeill LA, Riordan MV et al (2002) Hypoxia-inducible factor (HIF) asparagine hydroxylase is identical to factor inhibiting HIF (FIH) and is related to the cupin structural family. J Biol Chem 277:26351–26355

Higashi K, Clavo AC, Wahl RL (1993) In vitro assessment of 2-fluoro-2-deoxy-D-glucose, L-methionine and thymidine as agents to monitor the early response of a human adenocarcinoma cell line to radiotherapy. J Nucl Med 34:773–779

Higashi K, Ueda Y, Yagishita M et al (2000) FDG PET measurement of the proliferative potential of non-small cell lung cancer. J Nucl Med 41:85–92

Hughes CS, Shen JW, Subjeck JR (1989) Resistance to etoposide induced by three glucose-regulated stresses in Chinese hamster ovary cells. Cancer Res 49:4452–4454

Inoue T, Shibasaki T, Oriuchi N et al (1999) ^{18}F alpha-methyl tyrosine PET studies in patients with brain tumors. J Nucl Med 40:399–405

Ishiwata K, Hatazawa J, Kubota K et al (1989) Metabolic fate of L-[methyl-^{11}C]methionine in human plasma. Eur J Nucl Med 15:665–669

Ishiwata K, Kubota K, Kubota R et al (1991) Selective 2-[^{18}F]fluorodopa uptake for melanogenesis in murine metastatic melanoma. J Nucl Med 32:95–101

Ishiwata K, Hatazawa J, Kubota K et al (1996) A feasibility study on L-[1-carbon11]tyrosine and L-[methyl-carbon-11]methionine to assess liver protein synthesis. J Nucl Med 37:279–285

Isselbacher KJ (1972) Sugar and amino acid transport by cells in culture: differences between normal and malignant cells. N Engl J Med 286:929–933

Jager PL, Plaat BE, deVries EG et al (2000) Imaging of soft tissue tumors using L-3-[iodine-123]iodo-alpha-methyl-tyrosine SPECT: comparison with proliferative and mitotic activity, cellularity and vascularity. Clin Cancer Res 6:2252–2259

Molecular Imaging of Tumor Metabolism and Apoptosis

Jager PL, Vaalburg W, Pruim J et al (2001) Radiolabeled amino acids: basic aspects and clinical applications in oncology. J Nucl Med 42:432–445

Jansson T, Westlin JE, Ahlstrom H et al (1995) Positron emission tomography studies in patients with locally advanced and/or metastatic breast cancer: a method for early therapy evaluation. J Clin Oncol 13:1470–1477

Kaschten B, Stevenaert A, Sadzot B et al (1998) Preoperative evaluation of 54 gliomas by PET with fluorine-18-fluorodeoxyglucose and/or carbon-11-methionine. J Nucl Med 39:778–785

Kubota K, Matsuzawa T, Fujiwara T et al (1989) Differential diagnosis of AH109A tumor and inflammation by radioscintigraphy with L-[methyl-^{11}C]-methionine. Jpn J Cancer Res 80:778–782

Kubota K, Matsuzawa T, Fujiwara T et al (1990) Differential diagnosis of lung tumor with positron emission tomography: a prospective study. J Nucl Med 31:1927–1932

Kuwert T, Probst-Cousin S, Woesler B et al (1997) Iodine-123-alpha-methyl tyrosine in glioma: correlation with cellular density and proliferative activity. J Nucl Med 38:1551–1555

Lando D, Peet DJ, Gorman JJ, Whelan DA, Whitelaw ML, Bruick RK (2002a) FIH-1 is an asparaginyl hydroxylase enzyme that regulates the transcriptional activity of hypoxia-inducible factor. Genes Dev 16:1466–1471

Lando D, Peet DJ, Whelan DA, Gorman JJ, Whitelaw ML (2002b) Asparagine hydroxylation of the HIF transactivation domain a hypoxic switch. Science 295:858–861

Langen KJ, Coenen HH, Roosen N et al (1990) SPECT studies of brain tumors with L-3-[^{123}I] iodo-alpha-methyl tyrosine: comparison with PET, [124]IMT and first clinical results. J Nucl Med 31:281–286

Langen KJ, Ziemons K, Kiwit JC et al (1997) 3-[^{123}I]iodo-alpha-methyltyrosine and [methyl-^{11}C]-L-methionine uptake in cerebral gliomas: a comparative study using SPECT and PET. J Nucl Med 38:517–522

Lee ST, Scott AM (2007) Hypoxia positron emission tomography imaging with ^{18}F-Fluoromisonidazole. Semin Nucl Med 37:451–461

Leskinen-Kallio S, Lindholm P, Lapela M et al (1994) Imaging of head and neck tumors with positron emission tomography and [^{11}C]methionine. Int J Radiat Oncol Biol Phys 30:1195–1199

Lindholm P, Leskinen S, Lapela M (1998) Carbon-11-methionine uptake in squamous cell head and neck cancer. J Nucl Med 39:1393–1397

Martin SJ, Reutelingsperger CPM, McGahon AJ (1995) Early redistribution of plasma membrane phosphatidylserine is a general feature of apoptosis regardless of the initiating stimulus: inhibition by overexpression of Bcl-2 and Abl. J Exp Med 182:1545–1556

McConathy J, Martarello L, Malveaux EJ et al (2002) Radiolabeled amino acids for tumor imaging with PET: radiosynthesis and biological evaluation of 2-amino-3-[^{18}F]fluoro-2-methylpropanoic acid and 3-[^{18}F]fluoro-2-methyl-2-(methylamino)propanoic acid. J Med Chem 45:2240–2249

Mosskin M, Bergstrom M, Collins VP et al (1986) Positron emission tomography with 11C-methionine of intracranial tumors compared with histology of multiple biopsies. Acta Radiol 369 Suppl:157–160

Ogawa T, Kanno I, Shishido F et al (1991) Clinical value of PET with ^{18}F-fluorodeoxyglucose and L-methyl-^{11}C-methionine for diagnosis of recurrent brain tumor and radiation injury. Acta Radiol 32:197–202

Ogawa T, Shishido F, Kanno I et al (1993) Cerebral glioma: evaluation with methionine PET. Radiology 186:45–53

Padhani AR, Krohn KA, Lewis JS, Alber M (2007) Imaging oxygenation of human tumors. Eur Radiol 17:861–872

Plaat B, Kole A, Mastik M et al (1999) Protein synthesis rate measured with L-[1-^{11}C]tyrosine positron emission tomography correlated with mitotic activity and MIB-1 antibody-detected proliferation in human soft tissue sarcomas. Eur J Nucl Med 26:328–332

Plate KH, Breier G, Weich HA, Risau W (1992) Vascular endothelial growth factor is a potential tumor angiogenesis factor in human gliomas in vivo. Nature 359:845–848

Rau FC, Weber WA, Wester HJ et al (2002) O-(2-[^{18}F]fluoroethyl)-L-tyrosine (FET): a tracer for differentiation of tumor from inflammation in murine lymph nodes. Eur J Nucl Med 29:1039–1046

Reske SN, Kotzerke J (2001) FDG-PET for clinical use results of the 3rd German Interdisciplinary Consensus Conference, "Onko-PET III", 21 July and 19 September 2000. Eur J Nucl Med 28:1707–1723

Rodriguez M, Rehn S, Ahstrom H et al (1995) Predicting malignancy grade with PET in non-Hodgkin's lymphoma. J Nucl Med 36:1790–1796

Rozental JM, Levine RL, Nickles RJ, Dobkin JA (1989) Glucose uptake by gliomas after treatment. A positron emission tomographic study. Arch Neurol 46:1302–1307

Saier MH, Daniels JR, Boerner P, Lin J (1988) Animal amino acid transport systems in animal cells: potential targets of oncogene action and regulators of cellular growth. J Membr Biol 104:1–20

Scanlon K, Safirstein RL, Thies H et al (1983) Inhibition of amino acid transport by cis-diaminedichloroplatinum (II) derivatives L1210 murine leukemia cells. Cancer Res 43:4211–4215

Scanlon K, Cashmore AR, Kashani-Sabet M et al (1987) Inhibition of methionine uptake by methotrexate in mouse leukemia L1210. Cancer Chemother Pharmacol 19:21–24

Molecular Imaging of Tumor Metabolism and Apoptosis

Schaider H, Haberkorn U, Berger MR et al (1996) Application of alpha-amino-isobutyric acid, L-methionine, thymidine and 2-fluoro-2-deoxy-D-glucose to monitor effects of chemotherapy in a human colon carcinoma cell line. Eur J Nucl Med 23:55–60

Schwarzbach M, Willeke F, Dimitrakopoulou-Strauss A et al (1999) Functional imaging and detection of local recurrence in soft tissue sarcomas by positron emission tomography. Anticancer Res 19:1343–1349

Semenza GL (2002) HIF-1 and tumor progression: pathophysiology and therapeutics. Trends Mol Med 8:S62–S67

Shawver LK, Olson SA, White MK, Weber MJ (1987) Degradation and biosynthesis of the glucose transporter protein in chicken embryo fibroblasts transformed by the src oncogene. Mol Cell Biol 7:2112–2118

Shoup TM, Olson J, Hoffman JM et al (1999) Synthesis and evaluation of $[^{18}F]$1-amino-3-fluorocyclobutane-1-carboxylic acid to image brain tumors. J Nucl Med 40:331–338

Sieger S, Jiang S, Schönsiegel F, Eskerski H, Kübler W, Altmann A, Haberkorn U (2003) Tumor-specific activation of the sodium/iodide symporter gene under control of the glucose transporter gene 1 promoter (GTI-1.3). Eur J Nucl Med Mol Imaging 30:748–756

Sieger S, Jiang S, Kleinschmidt J, Eskerski H, Schönsiegel F, Altmann A, Mier W, Haberkorn U (2004) Tumor-specific gene expression using regulatory elements of the glucose transporter isoform 1 gene. Cancer Gene Ther 11:41–51

Singh D, Banerji AK, Dwarakanath BS, Tripathi RP, Gupta JP, Mathew TL, Ravindranath T, Jain V (2005) Optimizing cancer radiotherapy with 2-deoxy-d-glucose dose escalation studies in patients with glioblastoma multiforme. Strahlenther Onkol 181:507–514

Stadler P, Becker A, Feldmann HJ et al (1999) Influence of the hypoxic subvolume on the survival of patients with head and neck cancer. Int J Radiat Oncol Biol Phys 44:749–754

Teicher BA (2004) Hypoxia and drug resistance. Cancer Metastasis Rev 13:139–168

Uehara H, Miyagawa T, Tjuvajev J et al (1997) Imaging experimental brain tumors with 1-aminocyclopentane carboxylic acid and alpha-aminoisobutyric acid: comparison to fluorodeoxyglucose and diethylenetriaminepentaacetic acid in morphologically defined tumor regions. J Cereb Blood Flow Metab 17:1239–1253

Vaalburg W, Coenen HH, Crouzel C et al (1992) Amino acids for the measurement of protein synthesis in vivo by PET. Int J Rad Appl Instrum B 19:227–237

Vansteenkiste JF, Stroobants SG, Dupont PJ et al (1999) Prognostic importance of the standardized uptake value on (18)F-fluoro-2-deoxy-glucose-positron emission tomography scan in non-small-cell lung cancer: an analysis of 125 cases. Leuven Lung Cancer Group. J Clin Oncol 17:3201–3206

Villa P, Kaufmann SH, Earnshaw WC (1997) Caspases and caspase inhibitors. Trends Biochem Sci 22:388–393

Weber WA, Wester HJ, Herz M et al (2001) Kinetics of F-18-fluoroethyl-L-tyrosine (FET) in patients with brain tumors. J Nucl Med 42:214P–215P

Wertheimer E, Sasson S, Cerasi E, Ben-Neriah Y (1991) The ubiquitous glucose transporter GLUT-1 belongs to the glucose-regulated protein family of stress-inducible proteins. Proc Natl Acad Sci U S A 88:2525–2529

Wester HJ, Herz M, Weber W et al (1999) Synthesis and radiopharmacology of O-(2-[^{18}F]fluoroethyl)-L-tyrosine for tumor imaging. J Nucl Med 40:205–212

Widnell CC, Baldwin SA, Davies A, Martin S, Pasternak CA (1990) Cellular stress induces a redistribution of the glucose transporter. FASEB J 4:1634–1637

Wienhard K, Herholz K, Coenen HH et al (1991) Increased amino acid transport into brain tumors measured by PET of L-(2-^{18}F)fluorotyrosine. J Nucl Med 32:1338–1346

Willemsen AT, vanWaarde A, Paans AM et al (1995) In vivo protein synthesis rate determination in primary or recurrent brain tumors using L-[1-^{11}C]-tyrosine and PET. J Nucl Med 36:411–419

Wurker M, Herholz K, Voges J et al (1996) Glucose consumption and methionine uptake in low-grade gliomas after iodine-125 brachytherapy. Eur J Nucl Med 23:583–586

Young SD, Marshall RS, Hill RP (1988) Hypoxia induces DNA overreplication and enhances metastatic potential of murine tumor cells. Proc Natl Acad Sci U S A 85:9533–9537

Zwaal RFA, Schroit AJ (1997) Pathophysiologic implications of membrane phospholipid asymmetry in blood cells. Blood 89:1121–1132

Ernst Schering Foundation Symposium Proceedings, Vol. 4, pp. 153–187
DOI 10.1007/2789_2008_093
© Springer-Verlag Berlin Heidelberg
Published Online: 03 July 2008

Minimally Invasive Biomarkers for Therapy Monitoring

P. McSheehy[✉], P. Allegrini, S. Ametaby, M. Becquet,
T. Ebenhan, M. Honer, S. Ferretti, H. Lane, P. Schubiger,
C. Schnell, M. Stumm, J. Wood

Oncology Research, Novartis Pharma AG, 4002 Basel, Switzerland
email: *paul_mj.mcsheehy@novartis.com*

1	Introduction	154
2	Methods	155
2.1	Compounds	155
2.2	Animals	156
2.3	Interstitial Fluid Pressure	157
2.4	Magnetic Resonance Imaging	157
2.5	Positron Emission Tomography	158
2.6	Histology and Immunohistochemistry	159
2.7	Data Analysis	159
3	Results	160
3.1	Tumour IFP	160
3.2	Tumour IFP and MR-Measured Vascular Parameters	169
3.3	Effects of Compounds on FDG and FLT-PET	174
4	Conclusions	184
References		186

Abstract. Development of new drugs and optimal application of the drugs currently in use in clinical chemotherapy requires the application of biomarkers. Ideally, these biomarkers would stratify patients so that only those patients likely to respond to a particular therapy receive that therapy. However, that is not always feasible, and an alternative is to make use of early response biomarkers to determine the responding population. In this paper, a number of generic (i.e. not necessarily specific to the action mechanism of the compound) early-response biomarkers are discussed and compared in different models and with three compounds with quite different mechanisms of action: a VEGF-R inhibitor (PTK787), an mTOR inhibitor (RAD001) and a microtubule stabiliser (EPO906). The methods include noninvasive DCE-MRI and PET imaging for measuring tumour vascularity, metabolism and proliferation, as well as the minimally invasive WIN method for measuring tumour interstitial pressure (IFP). The data show that drug-induced changes in IFP (ΔIFP) involve mechanism-dependent changes in the tumour vascular architecture, and that ΔIFP may be considered a universal generic early-response marker of tumour response to therapy.

1   Introduction

Chemotherapy remains a mainstay for the treatment of cancer, which until 10 years ago involved predominantly cytotoxics, including alkylating agents, anti-metabolites and agents acting against microtubules. More recently, targeted agents have been added to the arsenal such as trastuzumab (Herceptin), imatinib (Glivec) and bevacizumab (Avastin). The successful development of the new generation of targeted agents demands the application of biomarkers so that the patient population is stratified, i.e. the right people get the right drugs. In fact, suitable biomarkers would aid the application of all chemotherapies since in some cancer indications response rates do not predict progression-free survival (PFS) or overall survival (OS). Furthermore, biomarkers are important for determining the optimal biological dose of a new compound in Phase-I/II trials, since these newer agents may not cause rapid changes in tumour size and thus RECIST (response-evaluation criteria in solid tumours), which was developed for the cytotoxics, may no longer be appropriate. Finally, generic biomarkers, i.e. markers not necessarily specific to the drug in question, can also be useful for measuring an early response of the tumour, before a change in tumour size (if any) or as an alternative or indeed better surrogate for some later endpoints.

A number of these generic biomarkers are now being used in the clinic to aid drug development of both traditional cytotoxics and the targeted agents. The approaches include monitoring blood biomarkers such as PSA or CA-125 as well as imaging biomarkers such as dynamic contrast-enhanced magnetic resonance imaging (DCE-MRI), positron-emission tomography using the glucose analogue [18]FDG (2′-[^{18}F]-fluoro-2′-deoxyglucose) or the thymidine analogue [18]FLT (3′-deoxy-3′-[^{18}F]-fluorothymidine) and the minimally invasive method of measuring tumour interstitial fluid pressure (IFP). Imaging methods are a powerful means of investigating fairly specific aspects of the tumour vasculature, metabolism and proliferation. Tumour IFP has for some time been identified as a parameter that in tumours is raised above normal tissue values, including human tumours in the clinic (Jain 1994), where it was identified as an independent prognostic indicator (Milosevic et al. 2001). Furthermore, recent clinical studies have shown that chemotherapy with bevacizumab (Willet et al. 2004) or paclitaxel (Taghian et al. 2005) can cause decreases in tumour IFP, which may be related to tumour response. Although the biological basis of a raised IFP is known to be related to a high vessel permeability, poor lymphatic drainage, poor perfusion and high cell density around the blood vessels as well as elevated glycolysis (Rutz 1999). The biology underlying a drug-induced change in IFP is less well studied.

The main aim of this paper was to focus on the three different early-response markers of MRI, PET and IFP and compare the response and biological data obtained for three different compounds currently in drug development, each with a different mechanism of action. In this way, the biological basis of a drug-induced change in IFP could be explored and the most appropriate early-response biomarker for each compound could be identified.

2 Methods

2.1 Compounds

Most studies utilised Novartis compounds synthesised in the Chemistry Department, Basel, Switzerland. This included the following compounds: NVP-AEE788, a dual inhibitor of VEGF-R and EGFR, EPO906 (generic name, patupilone), a microtubule stabiliser (MTS), PTK787 (generic name, vatalanib, aka PTK/ZK), a pan-VEGF-R inhibitor, RAD001 (generic name, everolimus), an mTOR inhibitor, STI571 (generic name, imatinib), a BCR-ABL, PDGF-R and c-Kit inhibitor, as well the alkylating agent cyclophosphamide (CP) and the anti-metabolite gemcitabine (Gemzar). Various pilot experiments were conducted to determine the optimal dose and schedule of the compounds to be used in vivo based upon achieving good anti-tumour efficacy and reasonable tolerability normally manifested as less than 15% body-weight loss.

Compounds were dosed daily (qd), weekly (qw) or two/three times per week (2qw, 3qw).

2.2   Animals

All animal studies conducted were in accordance with approved licenses as governed by the laws of the Kanton Basel.

2.2.1   Tumour Models

Human tumour xenografts were grown in Harlan nude mice (Novartis stock) following s.c. injection of cells in the animal flank. The human cell lines used were H-596 lung, HCT-116 colon, KB-31 cervical, HeLA cervical, U87MG glioma, 1A9 ovarian wild type and the paclitaxel-resistant form 1A9ptx10. In addition, two rat tumour xenografts in mice were created in the same way following s.c. injection of either C6 glioma or PROb colon cells. Tumour-bearing mice were used for efficacy, IFP and imaging studies once the tumour volume (TVol) was greater than 100 mm^3.

Syngeneic models used either rats or mice. The rat models were (a) A15 glioma and (b) PROb colon cells injected s.c. in the flank of BDIX rats and (c) BN472 breast tumour fragments implanted in the mammary fat-pad (orthotopic) of brown-Norway (BN) rats. Rats were normally in the range of 150–200 g body weight and tumours were used once the TVol was greater than 200 mm^3. The mouse models were (a) RIF-1 fibrosarcoma cells injected s.c. (effectively orthotopic) in the flank of C3H/He mice and (b) B16/BL6 melanoma cells injected under the skin (orthotopic) in the ear of C57/BL6 mice; the B16 tumours rapidly metastasise to the lymph nodes of the neck and the B16-mets were used for most studies. Both mouse models were used for studies after 10–14 days' growth. Mice were normally in the range of 20–25 g body

Minimally Invasive Biomarkers for Therapy Monitoring

weight. Further details of the models are available in Rudin et al. (2005) and Ferretti et al. (2005).

2.3 Interstitial Fluid Pressure

Tumour interstitial fluid pressure (IFP) was measured using the wick-in-needle (WIN) method in which a standard 23-gauge needle connected to a pressure transducer was inserted into the central part of the tumours, and the pressure monitored for a period of 10 min in animals anaesthetised with 2.5% isoflurane delivered at 2 l/min. (Ferretti et al. 2005).

2.4 Magnetic Resonance Imaging

Magnetic resonance imaging (MRI) was performed on both mouse B16/BL6 lymph-node metastases and rat BN472 mammary tumours. For all studies, animals were anaesthetised using 1.5% isoflurane in a 1:2 v/v mixture of O_2/N_2O and placed on an electrically warmed pad for cannulation of one of the tail-veins using a 30-guage needle attached to an infusion line of 30 cm and volume 80 μl to permit remote administration of the contrast agent. The animals were positioned on a cradle in a supine position inside the 30 cm horizontal bore magnet and the body temperature was maintained at $37\pm1°C$ using a warm air flow and was monitored with a rectal probe. MRI experiments were performed on a Bruker DBX 47/30 spectrometer at 4.7 T equipped with a self-shielded 12-cm bore gradient system, as previously described (Rudin et al. 2005).

2.4.1 Dynamic Contrast-Enhanced MRI

The low-molecular-weight (MW) contrast agent GdDOTA (gadolinium tetra-azocyclododecane-tetra-acetate, Dotarem®) was injected (0.1 mmol/kg) to determine tumour vascular permeability (initial slope of GdDOTA uptake, VP) and extravasion, i.e. tumour extracellular leakage space (final value for GdDOTA uptake, LS) and the first 20 points (102 s) from the injection of GdDOTA were used to calculate the initial area under the enhancement curve for the contrast agent (iAEUC). A similar approach was taken in some experiments using the high-

MW (>1 kD) contrast agent, Vistarem (P792, a dendrimeric Gd-based macromolecular contrast agent), which was injected at 200 μl/100 g body weight. Using an average arterial input function, K^{trans} estimates were obtained from a nonlinear regression analysis of the signal enhancement curves after Vistarem administration.

In some experiments, the iron oxide particle intravascular contrast agent, Endorem[®], was injected (6 mmol/kg of iron) 15–30 min after the other contrast agent, to determine the tumour relative blood volume (rBVol), and the initial slope of uptake of Endorem by the tumour was used as a blood flow index (BFI). The principles behind measurement of these parameters have already been fully described (Rudin et al. 2005).

2.4.2 Dynamic Susceptibility Contrast MRI

In rats bearing BN472 tumours, the contrast agent Sinerem, which is an iron oxide particle intravascular contrast agent, was injected i.v. (0.2 mmol/kg iron) for measurement of tumour vessel size and absolute tumour blood volume as a percentage of the total tumour size (aB-Vol). Briefly, the method is based on the measurement ΔR_2 and ΔR_2* relaxation rate constants induced by the injection of an intravascular slow-clearance superparamagnetic contrast agent. Based on relaxation theory, it was shown in vivo that the $\Delta R_2*/\Delta R_2$ ratio is related to the diameter of the vessels (Tropres et al. 2001).

2.5 Positron Emission Tomography

All experiments were conducted at the ETH as previously described (McSheehy et al. 2007). Briefly, a quad-HIDAC tomograph with a camera and four detector banks, each comprising four high-density avalanche chamber (HIDAC) modules, was used. The field of view was 280 mm axially and 170 mm in diameter, allowing the acquisition of whole-body images in a single bed position. Animals were anaesthetised with isoflurane in an air/oxygen mixture 20 min after injection of a maximum of 200 μl of the radiotracer [18]FDG (15–25 MBq per mouse) or [18]FLT (5–15 MBq per mouse) via a lateral tail vein. Acquisition of PET data was initiated 30 min after injection and lasted for 20–30 min. Data were acquired in list mode and reconstructed in a single time frame us-

Minimally Invasive Biomarkers for Therapy Monitoring

ing the OPL-EM algorithm with a bin size of 0.5 mm, a matrix size of 160×160×200 and a resolution recovery width of 1.3 mm. Reconstruction did not include scatter, random and attenuation correction.

Image files were evaluated by region-of-interest (ROI) analysis using the dedicated software Pmod. This provided total activity concentrations in the tumour (counts/ml), which were normalised to the injected dose per body weight (kBq/g) to give a unitless normalised uptake value (NUV).

2.6 Histology and Immunohistochemistry

At the endpoint of some experiments following efficacy, IFP or imaging studies, the nuclear staining dye, H33342, was injected i.v. at 20 mg/kg (2 mg/ml in normal saline) into anaesthetised animals and after 45 s, the animals were sacrificed by cervical dislocation and the tumour removed. The tumour was sliced through the central plain and analyzed using a magnification of 100x to determine the blood vessel width and blood vessel density (Rudin et al. 2005).

In other experiments, nothing was injected and tumour slices were harvested from the largest circumference of the tumour and fixed in 4% phosphate buffered formaldehyde for 24 h at 4°C and processed into paraffin before preparation and staining by (a) haematoxylin and eosin (H&E) for evaluation of necrosis, (b) caspase-3 to measure apoptosis and (c) Ki67 to measure proliferation, as previously described (Ferretti et al. 2005).

2.7 Data Analysis

2.7.1 Efficacy and Tolerability

Body weight, BW (g) and tumour volume (TVol) were measured three times per week. Efficacy was assessed as the T/C from the change (difference) in TVol using the endpoint (at 1–3 weeks after treatment initiation) to give a T/C_{TVol}. In a similar manner, the effect of the drug on BW was also quantified to give the fractional change in BW (ΔBW), i.e. BW_{after}/BW_{before}, for both vehicle-control and drug-treated groups so that the $T/C_{BW} = \Delta BW_{(treated)}/\Delta BW_{(control)}$.

2.7.2 Biomarkers: IFP, MR, PET and IHC

Individual values for each tumour at each different time point from before treatment (day 0) and after treatment (1–10 days after treatment initiation) are presented. Alternatively, or in addition, the fractional change (ΔF) in the parameter (final value/starting value) for each tumour at each time point after treatment is also presented from which the T/C for that parameter may be determined as described for BW, e.g. T/C_{IFP} or T/C_{BVol} or T/C_{FLT} etc.

2.7.3 Statistics

Differences between the means of TVol and BW of vehicle or compound-treated animals were assessed using a two-tailed t-test or one-way ANOVA, as required. If the normality and/or the equal variance test failed, a logarithmic transformation was applied; however, this did not always normalise the data.

For the biomarker analysis, data was examined in two different ways: (a) using the raw values and a two-way repeated measures analysis of variance (2W-RM-ANOVA) with Tukey post-hoc and (b) comparing the differences in fractional changes of the mean values induced by treatments using a one-way ANOVA with Tukey-test post-hoc or a two-tailed t-test, as appropriate.

All data are shown as mean±SEM except where stated to be ±SD, with the significance value set at $p < 0.05$. In all the figures, significance is represented as follows: * $p < 0.05$, ** $p < 0.01$, *** $p < 0.001$.

3 Results

3.1 Tumour IFP

We evaluated basal tumour IFP (mm Hg) in ten different experimental tumour models, including human tumours in nude mice as well as syngeneic models in mice and rats (Table 1). The highest mean IFPs were all associated with rat tumours, whether grown in mice or rats s.c. or orthotopically. However, rat colon PROb tumours implanted s.c. (ectopically) in rats exhibited an IFP of 15.2±0.7 mm Hg, ($n = 45$), whereas the

Minimally Invasive Biomarkers for Therapy Monitoring

Table 1 Basal tumour IFP in different experimental models

Tumour	Histotype	Species	Host, Strain	Type	n	IFP
PROb	Colon	Rat	Rat, BDIX	S.c.	45	15.2 ± 4.7
C6	Glioma	Rat	Nude mouse, Harlan	S.c.	36	14.6 ± 9.0
C6	Glioma	Rat	Nude mouse, Balb/c	S.c.	35	13.7 ± 9.5
BN472	Breast	Rat	Rat, BN	S.c.	12	13.1 ± 5.4
BN472	Breast	Rat	Rat, BN	Ortho	218	12.3 ± 4.4
HeLa	Cervical	Human	Nude mouse, Harlan	S.c.	15	11.2 ± 5.8
PROb	Colon	Rat	Nude mouse, Harlan	S.c.	14	10.5 ± 8.6
1A9	Ovarian	Human	Nude mouse, Harlan	S.c.	62	7.2 ± 4.7
1A9ptx10	Ovarian	Human	Nude mouse, Harlan	S.c.	65	6.2 ± 4.0
B16/BL6	Melanoma	Mouse	Mouse, C57/BL6	Ortho	35	6.7 ± 3.0
B16/BL6	Melanoma	Mouse	Mouse, C57/BL6	S.c.	14	6.3 ± 3.3
RIF-1	Fibrosarc	Mouse	Mouse, C3H/He	S.c./orth	29	6.3 ± 3.2
U87MG	Glioma	Human	Nude mouse, Harlan	S.c.	24	4.4 ± 3.4

The results show the mean $\pm$ SD IFP in mmHg. Tumours were implanted either ectopically (s.c.) or orthotopically (ortho)

same tumour also implanted s.c. in the flank of nude mice had a significantly ($p = 0.01$) lower IFP (10.5 ± 2.3 mm Hg, $n = 14$). This observation emphasises the important role of the host in the growth of a tumour and thus also the basal IFP of that tumour. In contrast, the IFP of C6 tumours was not significantly altered when grown s.c. in two different types of nude mouse (Harlan or Balb/c). Furthermore, in two syngeneic orthotopic models (BN472 in rats and B16/BL6 in mice), there was no evidence that the implantation site was important (Table 1). In the BN472 and B16 models, tumours were in some cases grown both s.c. and orthotopically in the same animal allowing a paired comparison which

Table 2 Examples of the effect of different compound treatments on tumour IFP in different tumour models

Compound	Dose and schedule	Day 2/3		Day 6/7	
		T/C_{TVol}	T/C_{IFP}	T/C_{TVol}	T/C_{IFP}
BN472 rat breast model					
AEE788	60 mg/kg, 3 qw	0.44	0.48*	0.48*	0.49
PTK787	200 mg/kg, qd	0.47	0.66	0.3	0.5*
RAD001	5 mg/kg, 3 qw	0.96	0.74	0.75	0.62*
EPO906	1.5 mg/kg, once	0.36	0.52*	0.06*	0.32*
STI571	100 mg/kg, 2 qd	0.04*	0.75*	–	–
CP	50 mg/kg, qw	0.01*	0.82	−0.2*	0.65*
Gemzar	150 mg/kg, qw	0.07	0.51*	0.07*	1.18
1A9 human ovarian model					
EPO906	4 mg/kg, once	0.30	0.47*	0.14*	0.18*
PROb rat colon model					
STI571	200 mg/kg, qd	0.54	0.59*	–	–

The results show the drug-induced change TVol (*efficacy*) and the IFP in a one or two (*pooled*) experiments ($n = 6$–21) with respect to vehicle-treated animals as the respective T/C (see Methods), where *indicates a significant difference ($p < 0.05$) to vehicle

showed no significant difference and thus indicating that implantation-site had no effect on IFP ($P < 0.7$). In general, most human tumour s.c. xenografts had rather low IFPs and in some cases (e.g. U87MG), there were several tumours that were almost unmeasurable (<2.0 mm Hg). One other paired comparison was possible: human ovarian tumours 1A9 (wild type) and the paclitaxel-resistant tumour (1A9ptx10), which has

Minimally Invasive Biomarkers for Therapy Monitoring   163

mutations in the drug's molecular target of β-tubulin; here there was also no significant difference in IFP.

3.1.1 Effect of Different Compounds on Tumour IFP

All compounds tested, independent of the tumour type and their mechanism of action, reduced tumour IFP significantly either fairly rapidly (after 2–3 days) or after 1 week (Table 2). The largest decreases in IFP were associated with the microtubule inhibitor EPO906, typically greater than 70% decreases in IFP at maximal doses, and an example of a longitudinal experiment in the RIF-1 model with EPO906 is shown in Fig. 1. For all compounds tested, the decrease in IFP was dose-dependent and there was a plateau in the maximum decrease achievable. Significant decreases in tumour IFP only occurred when there was a significant decrease in tumour volume (TVol) caused by compound treatment. This suggested there was a relationship between decreases in IFP and the effectiveness of anti-tumour efficacy. In fact, highly significant positive correlations could be shown between early (day 2/3) decreases in IFP (expressed as the fractional change for each tumour, ΔIFP) and the eventual change (i.e. ΔTVol) of that tumour (see Fig. 2). Again, the strength of the correlation was independent of the mechanism of action of the compound, with the targeted agents PTK787, EPO906 and gemcitabine all showing similar r-values (0.61–0.68; $p < 0.01$). However, these correlations were only significant when the studies were done on orthotopic tumours (BN472, RIF-1); other studies using the 1A9ptx10, PROb, HeLa or the U87MG models failed to reach significance except for 1A9wt, but then only after 7 days (Table 3).

3.1.2 Basal IFP and Compound Response

In general, the basal IFP of tumours showed no correlation with the outcome of tumour therapy. In part this may reflect that experimental tumour models are designed to show minimal heterogeneity and also that many of the studies from one dose were relatively small even when pooling different experiments. Nevertheless, in two cases in the BN472 model, PTK787 ($n = 19$) and EPO906 ($n = 11$) showed trends for a significant ($p = 0.04$ to 0.07) negative correlation between the basal IFP and

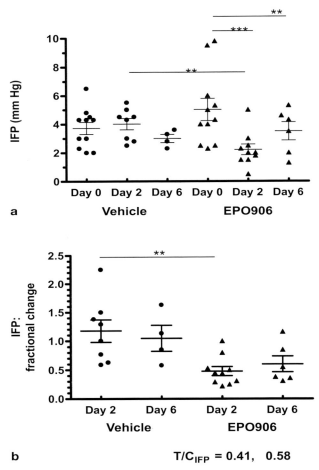

Fig. 1a,b. Longitudinal effect of EPO906 on the IFP of RIF-1 tumours. C3H mice bearing RIF-1 tumours were treated with EPO906 (5 mg/kg i.v.) or vehicle once on day 0 and tumour IFP measured using the WIN method. The results show (**a**) IFP values for individual tumours and associated mean±SEM on respective days (significance assessed using a 2W-RM-ANOVA), (**b**) individual fractional changes in IFP and associated mean±SEM relative to baseline measurements (significance assessed using a two-tailed t-test)

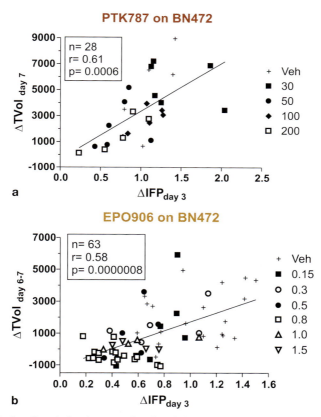

Fig. 2a,b. Correlation between fractional change in IFP and efficacy. The results show the compound-induced fractional change in IFP (ΔIFP) 2–3 days after treatment initiation and the tumour volume change (ΔTVol) after 6–7 days for individual tumours treated with different compound doses (mg/kg) shown on the *right-hand side* of the graphs. PTK787 treatment was daily, EPO906 and Gemzar weekly (q.w.) or twice weekly (2 q.w.)

TVol on day 7 (Fig. 3). Since a smaller change in TVol reflects a better response, this data implies that tumours with a higher IFP may actually

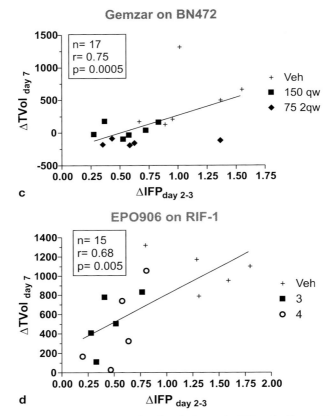

Fig. 2c,d. Correlation between fractional change in IFP and efficacy. The results show the compound-induced fractional change in IFP (ΔIFP) 2–3 days after treatment initiation and the tumour volume change (ΔTVol) after 6–7 days for individual tumours treated with different compound doses (mg/kg) shown on the *right-hand side* of the graphs. PTK787 treatment was daily, EPO906 and Gemzar weekly (q.w.) or twice weekly (2 q.w.)

be expected to respond better to drug treatment, and thus measurement of the basal IFP of tumours might also aid patient stratification.

Table 3 Summary of correlations between compound-induced fractional changes in IFP and efficacy

Tumour model	Host	Drug	ΔIFP$_{day\ 2-3}$ vs ΔTVol$_{day\ 6-7}$			ΔIFP$_{day\ 6-7}$ vs ΔTVol$_{day\ 6-7}$		
			n	r	P	n	r	P
PROb	Nude mice	STI571	14	0.30	0.30	14	0.32	0.26
PROb	BDIX rat	STI571	12	0.22	0.49	12	0.27	0.40
BN472	*BN rat*	*AEE788*	*26*	*0.11*	*0.60*	*26*	*0.07*	*0.72*
BN472	*BN rat*	*PTK787*	*28*	***0.61***	***0.0006***	*28*	***0.64***	***0.0002***
BN472	*BN rat*	*RAD001*	*14*	*0.09*	*0.77*	*14*	*0.30*	*0.30*
1A9	Nude mice	EPO906	50	0.19	0.19	50	**0.31**	**0.005**
1A9ptx10	Nude mice	EPO906	58	0.23	0.08	58	0.11	0.40
U87MG	Nude mice	EPO906	22	0.04	0.85	22	0.02	0.95
HELA	Nude mice	EPO906	15	0.26	0.35	15	0.03	0.92
RIF-1	*C3H/He mice*	*EPO906*	*14*	***0.68***	***0.005***	*14*	*0.30*	*0.28*
BN472	*BN rat*	*EPO906*	*64*	***0.48***	***0.00005***	*64*	***0.43***	***0.0004***
RIF-1	*C3H/He mice*	*CP*	*11*	*0.33*	*0.32*	*11*	*0.19*	*0.58*
BN472	*BN rat*	*CP*	*16*	*0.36*	*0.17*	*16*	***0.64***	***0.008***
BN472	*BN rat*	*Gemzar*	*17*	***0.75***	***0.0005***	*17*	*0.03*	*0.92*

The results show the Pearson correlation coefficient, *r*, and the associated *p*-value. Values in **bold** indicate significant relationships; italic indicates orthotopic models

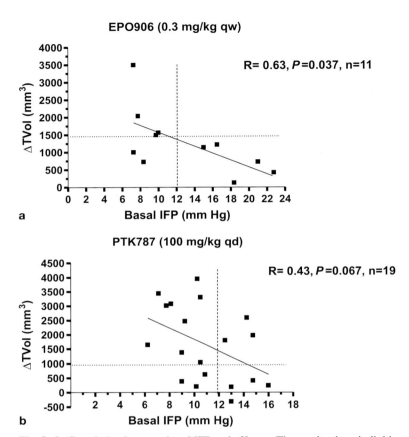

Fig. 3a,b. Correlation between basal IFP and efficacy. The results show individual BN472 tumour IFP values before treatment and correlation with the eventual change in TVol after 7 days for (**a**) EPO906 (**b**) PTK787

Fig. 4a,b. Longitudinal effect of EPO906 on the IFP and BVol of BN472 tumours. Rats bearing BN472 tumours were treated with EPO906 (0.8 mg/kg i.v.) or vehicle once on day 0 and tumour IFP and aBVol were measured as described in Methods. The results show the individual fractional changes in IFP and aBVol relative to baseline measurements (significance assessed using a two-tailed t-test)

3.2 Tumour IFP and MR-Measured Vascular Parameters

3.2.1 Comparison of Effects of EPO906 and PTK787

The effects of these two compounds on tumour growth, IFP and MR-measured vascular parameters were investigated in two different mod-

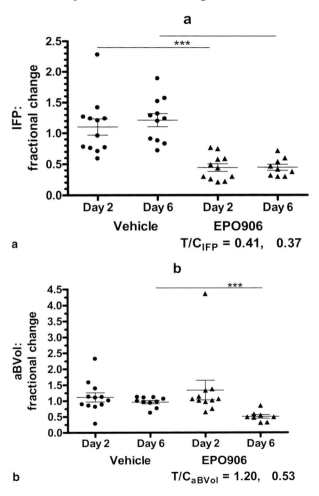

els. In the murine B16/BL6 orthotopic melanoma model, both compounds at optimal tolerable doses were able to significantly reduce tumour growth and caused significant decreases in IFP 2–4 days after treatment initiation. For EPO906, this was paralleled by a significant decrease in tumour blood volume (rBVol), with no change in blood vessel permeability (VP), intracellular leakage space (LS) or the composite parameter of iAUEC for the contrast agent Dotarem (Ferretti et al. 2005). For PTK787, VP and especially LS and iAUEC were reduced while the intravascular contrast agent, Endorem, showed that tumour blood flow (BFI) increased and BVol was not significantly affected (Rudin et al. 2005; Lee et al. 2006). To a certain extent, these results were consistent with the known mechanism of action of the two compounds; this will be discussed further below. However, IFP and the MRI-measured vascular parameters were not determined in the same animals, so the relationship of these parameters was not clear.

The effects of both PTK787 and EPO906 were investigated more extensively using the rat syngeneic orthotopic model of BN472 breast tumours.

EPO906 (0.8 mg/kg i.v. once) decreased significantly both IFP and BVol (Fig. 4a, b) and also tended to reduce the mean vessel size by day 6 ($T/C = 0.37$, $p = 0.15$). After 6 days, there was a significant inhibition of tumour growth, actually regression ($T/C_{TVol} = -0.3$), and histological analysis ex vivo demonstrated a 2.4-fold increase in apoptosis with respect to vehicle-treated rats but no change in necrosis. IFP and aBVol were strongly positively correlated with each other on day 6, and also with apoptosis and TVol (Fig. 5). Other experiments (data not shown) showed that the epothilone could significantly increase apoptosis in the BN472 tumour model and RIF-1 tumours after just 2 days of treatment. Thus, although both BVol and apoptosis were strongly correlated with IFP on day 6, the changes in BVol were slower than those in IFP, suggesting that at least at early time points, decreases in IFP may reflect more tumour cell death than destruction of the vasculature. Finally, further experiments in this model using either Dotarem or the higher-MW contrast agent Vistarem also failed to demonstrate significant decreases in K^{trans} either at day 2 or day 7 (maximum $T/C_{Ktrans} = 0.9$ and 0.8, respectively with Vistarem), confirming the lack of effect of EPO906 on vascular permeability observed in the B16/BL6 model.

In the BN472 model, PTK787 (100 mg/kg p.o. daily) significantly decreased tumour IFP by 25% and increased BFI by 40% after 3 days of treatment (Fig. 6a, b). However, changes were rather variable and significant changes could not be detected in blood vessel permeability, although there were strong trends indicating a decrease in leakage space and the composite parameter of iAUEC, as well as vessel width and rBVol (data not shown). Histological analysis also tended to show a decrease in the mean blood vessel width from 20.8±1.3 μm ($n = 14$) to 17.7±1.0 μm ($n = 15$) ($p = 0.068$) but not blood vessel density (14–16 vessels/mm^2). In the experiments where both IFP and MR

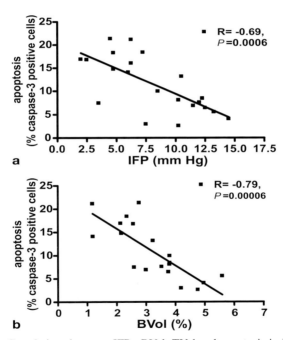

Fig. 5a,b. Correlations between IFP, aBVol, TVol and apoptosis in BN472 tumours following EPO906 treatment. Rats bearing BN472 tumours were treated with EPO906 (0.8 mg/kg); apoptosis, IFP, aBVol and TVol were measured on day 6. Data show the Pearson correlation coefficient and associated *p*-value

parameters were measured in the same rat tumour, there was a significant correlation between the pretreatment IFP and the BVol and BFI ($r = 0.45$, $p < 0.05$). Furthermore, as for EPO906, the change in IFP correlated with the change in BVol ($r = 0.71$, $p < 0.01$), although this did not reach significance for IFP and BFI. Thus, decreases in IFP in the BN472 tumour induced by PTK787 were associated with increased blood flow and narrower blood vessels rather than decreased blood volume and narrower blood vessels as for EPO906. The effects on tumour blood flow may have masked the decrease in vascular permeability that PTK787 would be expected to cause, since transfer of the low-MW contrast agent, Dotarem (GdDTPA), across the vasculature is influenced by both permeability and flow. Indeed, subsequent ex-

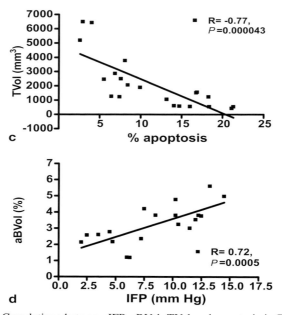

Fig. 5c,d. Correlations between IFP, aBVol, TVol and apoptosis in BN472 tumours following EPO906 treatment. Rats bearing BN472 tumours were treated with EPO906 (0.8 mg/kg); apoptosis, IFP, aBVol and TVol were measured on day 6. Data show the Pearson correlation coefficient and associated *p*-value

Minimally Invasive Biomarkers for Therapy Monitoring

periments using a higher-MW contrast agent (Vistarem) in the BN472 model were able to demonstrate marked (T/C_{Ktrans} = 0.37) and highly significant ($p < 0.01$) decreases in the vascular transfer constant, K^{trans} after only 2 days of treatment (Schnell et al. 2008).

3.2.2 Effects of RAD001 on MR-Measured Parameters

RAD001 at 10 mg/kg p.o. daily was an effective inhibitor of growth, if not highly potent, of both B16/BL6 and BN472 tumours, leading to T/C_{TVol} of 0.5 and 0.63, respectively. As discussed above, RAD001 also caused a significant decrease in the IFP of BN472 tumours, but this was a late (day 7) rather than early effect (day 2/3). The effect of RAD001 on the IFP of B16/BL6 tumours was not investigated. RAD001 did not significantly affect any MR-measured vascular parameter after 2–3 days of treatment in either tumour model. There was a trend for the iAUEC (Dotarem or Vistarem) and K^{trans} (Vistarem) to increase, and in one experiment in B16/BL6 tumours, K^{trans} was significantly increased relative to vehicle (T/C_{Ktrans} = 2.3) after 6 days, while BVol tended to decrease and BFI to increase. No experiments were conducted in which both IFP and MR-measured vascular parameters were determined. Nevertheless, the data for RAD001 suggest that decreases in IFP were slow, but as for PTK787, eventually reflected increases in tumour blood flow manifested as a raised BFI and K^{trans}.

3.2.3 IFP and Tumour Vasculature: Conclusions

The data presented support the concept of a relationship between tumour IFP and tumour blood flow, permeability and blood volume. Although all compounds tested could reduce IFP either rapidly or after several days, the underlying reasons for the change in IFP depend upon the nature of the compound's mechanism of action. Thus, the classic anti-angiogenic agent, PTK787, which interferes with VEGF-R signalling on endothelial cells, reduced K^{trans}, narrowed the blood vessels and increased blood flow, events consistent with a normalisation of the tumour vasculature. The mTOR inhibitor RAD001, which inhibits both tumour cell and endothelial cell proliferation, slowly increased K^{trans} and blood flow. The MTS, EPO906, did not affect vascular permeabil-

174 P. McSheehy et al.

ity, but did rapidly reduce blood volume, consistent with a strong anti-vascular effect.

3.3 Effects of Compounds on FDG and FLT-PET

3.3.1 In vivo Tumour Uptake of [18]FDG

Uptake and retention of the glucose analogue, [18]FDG, is considered to reflect both tumour cell glycolysis and cell viability but may also be influenced by other nonspecific factors including blood flow and infiltration by inflammatory cells such as macrophages (Mankoff et al. 2003). Indeed, our FDG-PET studies have shown that in the poorly vascularised s.c. xenograft models, the FDG signal is rim-limited while in well-vascularised orthotopic models (BN472, B16/BL6, RIF-1 and others) the signal is more homogeneous across the tumour.

PTK787 dosed at 100 mg/kg p.o. daily, failed to significantly affect FDG uptake by BN472 tumours after either 2 or 7 days of treatment (Fig. 7a, b).

In contrast, EPO906 (0.8 mg/kg i.v. once) in the same model, significantly reduced the FDG NUV at both time points, with the greatest effect being apparent after just 2 days, after which there was some recovery (Fig. 7c, d).

RAD001 (10 mg/kg p.o. daily) was tested in several different mouse models using either human tumour xenografts implanted in nude mice or the syngeneic orthotopic B16/BL6 model (McSheehy et al. 2007). The models used could be divided into two types based upon their sensitivity to inhibition of proliferation by RAD001 in vitro, i.e. the IC50. Thus, the human lung H-596 and murine melanoma B16/BL6 models had low IC50s of 5 nM and 0.7 nM, respectively, while the human cervical KB-31 and human colon HCT-116 had high IC50s of 1.8 and

Fig. 6a,b. Effect of PTK787 on the IFP and BFI of BN472 tumours. Rats bearing BN472 tumours were treated with PTK787 (100 mg/kg p.o. daily) or vehicle; tumour IFP and BFI were measured as described in methods on day 0 and day 3. The results show the individual fractional changes in IFP and BFI relative to baseline measurements (significance assessed using a two-tailed t-test)

Minimally Invasive Biomarkers for Therapy Monitoring

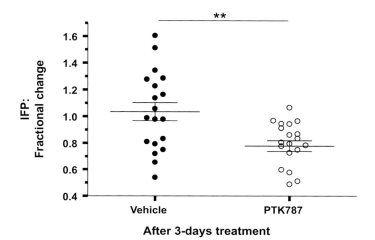

a $T/C_{IFP} = 0.75$

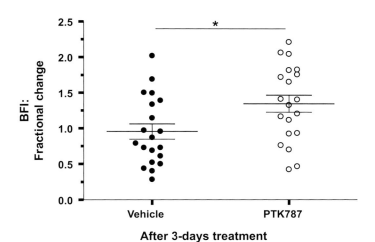

b $T/C_{BFI} = 1.42$

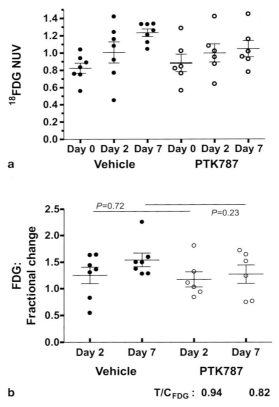

Fig. 7a,b. Effect of PTK787 or EPO906 on FDG uptake by BN472 tumours. Rats bearing BN472 tumours were treated with PTK787 (100 mg/kg p.o. daily) or EPO906 (0.8 mg/kg once) and the FDG uptake semi-quantified as the normalised uptake value (*NUV*) measured on days 0, 2, 6 or 7. The results show the values for individual tumours (a,c) and associated mean±SEM on respective days (significance assessed using a 2W-RM-ANOVA) and individual fractional changes (b,d) and associated mean±SEM relative to baseline measurements (significance assessed using a two-tailed *t*-test)

4.2 µM, respectively. A single oral dose of RAD001 leads to a Cmax in tumour-bearing mice of approximately 100 nM with a half-life in

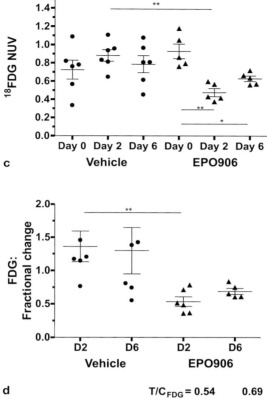

Fig. 7c,d. Effect of PTK787 or EPO906 on FDG uptake by BN472 tumours. Rats bearing BN472 tumours were treated with PTK787 (100 mg/kg p.o. daily) or EPO906 (0.8 mg/kg) and the FDG uptake semi-quantified as the normalised uptake value (*NUV*) measured on days 0, 2, 6 or 7. The results show the values for individual tumours (a,c) and associated mean±SEM on respective days (significance assessed using a 2W-RM-ANOVA) and individual fractional changes (b,d) and associated mean±SEM relative to baseline measurements (significance assessed using a two-tailed *t*-test)

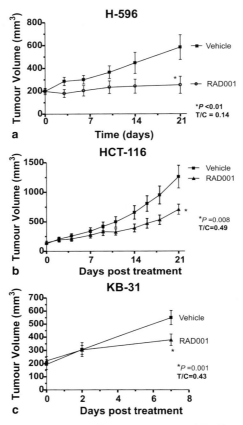

Fig. 8. Efficacy of RAD001 in different tumour models. Tumours were created as described in "Methods", and once they reached approximately 200 mm^3 they were treated daily with RAD001 (10 mg/kg p.o.). The results show the mean±SEM

the tumour and plasma of 8 and 16 h, respectively. Thus, the KB-31 and HCT-116 models would not be expected to respond to RAD001 in vivo, assuming the tumour cells were the only target. However, as Fig. 8 shows, after 1 week or more of treatment, significant inhibition

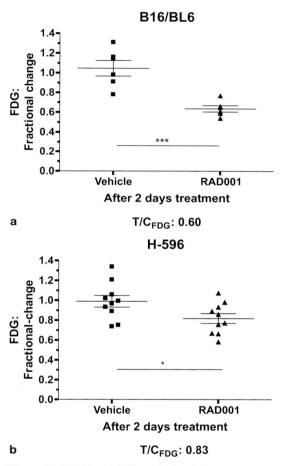

Fig. 9a,b. Effect of RAD001 on FDG uptake by different tumour models. The results show the fractional change in FDG-NUV after 2 or 7 days of daily treatment (10 mg/kg p.o. daily) and the associated T/C. The results show the values for individual tumours and associated mean±SEM on respective days (significance assessed using a 2W-RM-ANOVA) and individual fractional changes and associated mean±SEM relative to baseline measurements (significance assessed using a two-tailed *t*-test)

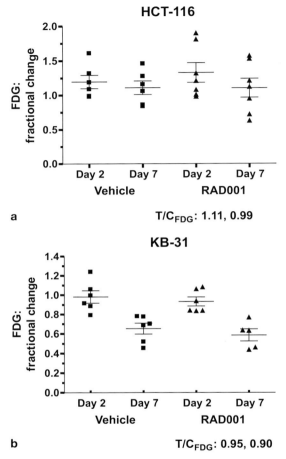

Fig. 9c,d. Effect of RAD001 on FDG uptake by different tumour models. The results show the fractional change in FDG-NUV after 2 or 7 days of daily treatment (10 mg/kg p.o. daily) and the associated T/C. The results show the values for individual tumours and associated mean±SEM on respective days (significance assessed using a 2W-RM-ANOVA) and individual fractional changes and associated mean±SEM relative to baseline measurements (significance assessed using a two-tailed t-test)

Fig. 10a–c. Effect of RAD001 on FLT uptake in two different tumour models and comparison with IHC measurements ex vivo. The results show the fractional change in FLT-NUV after 2–10 days daily treatment (10 mg/kg p.o. daily) and the associated T/C. In addition, for each model the IHC analysis of the Ki67 and apoptotic index is shown for each treatment group. For FLT, the results show the values for individual tumours and associated mean±SEM on respective days (significance assessed using a 2W-RM-ANOVA) and individual fractional changes and associated mean±SEM relative to baseline measurements (significance assessed using a two-tailed t-test). For IHC, the values for individual tumours and associated mean±SEM are shown for the respective endpoint

of growth, albeit modest, could be achieved, presumably reflecting the anti-angiogenic/anti-vascular activity of the compound. In confirmation of this hypothesis, significant ($p < 0.001$) decreases in blood vessel density were seen in the HCT-116 model: 4.8 ± 0.5 vessels/mm^2 ($n = 10$), reduced to 2.6 ± 0.6 vessels/mm^2 ($n = 10$) after 3 weeks of RAD001 treatment. Despite these growth-inhibitory effects, no significant changes in FDG NUV were apparent in either of the less sensitive models after 2–7 days of treatment, while in contrast, the more sensitive models showed significant decreases after just 2 days of treatment (Fig. 9).

3.3.2 In vivo Tumour Uptake of ^{18}FLT

EPO906 (5 mg/kg i.v. bolus) caused a rapid decrease in the FLT-NUV of murine RIF-1 tumours (T/C$_{FLT} = 0.78$) and this was associated, as expected, with a similar decrease in Ki67, which correlated significantly with the FLT-change after 24 h ($r = 0.42$, $p = 0.02$).

Daily RAD001 treatment (10 mg/kg p.o.) for 2 days caused a small but highly reproducible decrease in the ^{18}FLT-NUV of H-596 tumours (T/C$_{FLT} = 0.87$, 0.82, 0.80 for three independent experiments) but did not alter uptake in HCT-116 tumours even after 10 days of treatment (Fig. 10). IHC analysis showed that in neither of these models was there any significant change in apoptosis or Ki67.

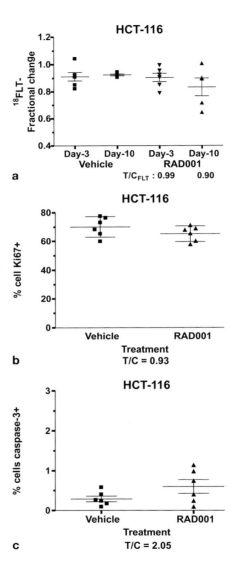

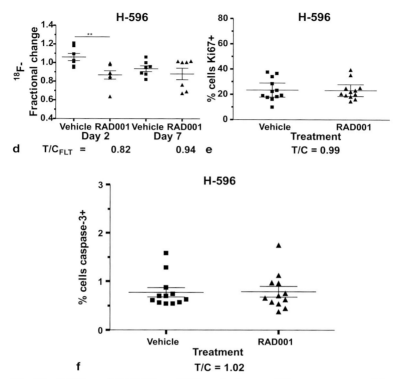

Fig. 10e–f. Effect of RAD001 on FLT uptake in two different tumour models and comparison with IHC measurements ex vivo. The results show the fractional change in FLT-NUV after 2–10 days daily treatment (10 mg/kg p.o. daily) and the associated T/C. In addition, for each model the IHC analysis of the Ki67 and apoptotic index is shown for each treatment group. For FLT, the results show the values for individual tumours and associated mean±SEM on respective days (significance assessed using a 2W-RM-ANOVA) and individual fractional changes and associated mean±SEM relative to baseline measurements (significance assessed using a two-tailed *t*-test). For IHC, the values for individual tumours and associated mean±SEM are shown for the respective endpoint

4 Conclusions

The basal IFP of a tumour is dependent both upon the individual tumour cells and the milieu in which it grows so that the type of host (e.g. mouse vs rat) can strongly affect the basal IFP, but not the site within an individual host. If one can draw a further conclusion from this relatively small study it would be that rat tumour cells in rats tend to have higher IFPs than mouse or human tumour cells grown in mice. With regard to drug response, there was some evidence that the basal IFP could also be a stratifier, since tumours with a higher IFP tended to show the best response to the compounds EPO906 and PTK787. However, these were not highly significant correlations (approximately $p = 0.05$), especially in comparison to the correlations found for the early fractional change in IFP (ΔIFP) and the eventual change in tumour volume (ΔTVol) for the compounds EPO906, Gemzar and PTK787 in the two orthotopic models of BN472 and RIF-1. Such data suggest that IFP changes could be a useful generic early-response marker for tumour response to therapy. However, these significant correlations were not evident in the ectopic models, except in the 1A9 model treated with EPO906, where there was a weaker correlation ($r = 0.3$, $p < 0.01$) but only when comparing late ΔIFP and ΔTVol. This may well reflect the different vascular architecture of ectopic and orthotopic tumours, as illustrated by a number of methods discussed in this report. DCE-MRI showed that the ectopic models were much better perfused on the rim than in the large, often necrotic portion of the tumour, and FDG-PET also showed very strikingly that viable cells and/or blood flow were rim-limited; of course this was confirmed ex vivo by H&E histology. In general, this suggests that only the well-vascularised tumour models clearly demonstrated a correlation between an early decrease in IFP and the eventual tumour response.

Consistent with the above hypothesis were the results obtained with EPO906 and PTK787, which showed that both basal IFP and the drug-induced changes in IFP were related to vascular permeability, blood flow and total blood volume. The precise relationship depended on the compound's mechanism of action. Thus, the cytotoxic anti-vascular agent caused large decreases in BVol and also appeared to reduce the mean vessel size, without significantly affecting vascular permeability or blood flow, while the VEGF-R targeted agent reduced permeability and increased blood flow without affecting blood volume. Histological analysis also confirmed a trend showing reduced mean vessel width. Both compounds therefore caused the tumours to manifest examples of a normalisation of the vasculature, albeit in different ways, with the anti-angiogenic activity being associated, perhaps transiently, with an increase in blood flow, thus masking the expected primary effect of reducing vascular permeability.

The data obtained in the FDG-PET experiments with PTK787 and RAD001 could also be considered to support the notion that anti-angiogenic activity is associated with an increase in tumour blood flow. Despite significantly inhibiting BN472 tumour growth, PTK787 failed to significantly impact FDG uptake after 2 or 7 days, and a similar phenomenon was seen with RAD001 in the less sensitive tumour models where the growth-inhibitory effect of RAD001 is considered to be at the level of the endothelium. The reason for this may be that a drug-induced increase in blood flow could mask a change in the number of viable actively glycolytic cells, since the method could not distinguish between FDG (intra- or extracellular) or phosphorylated FDG. Indeed, EPO906, which did not affect tumour blood flow, caused a large and rapid decrease in FDG uptake by BN472 tumours. In sensitive models, RAD001 was also able to significantly decrease tumour FDG uptake.

To a certain extent, similar data have so far been obtained using the PET-proliferation marker of FLT. RAD001 caused small but consistent decreases in FLT uptake by H-596 tumours, but not in the less sensitive HCT-116 tumour, a result similar to that seen with FDG-PET; and EPO906 caused a rapid decrease in FLT uptake of RIF-1 tumours.

Thus to sum up. Tumour IFP, and especially drug-induced changes in IFP, reflect the tumour vascular architecture and may be used as a generic early marker of tumour response. In all models studied, significant tumour efficacy is always eventually associated with a decrease in IFP. Interestingly, the same cannot be said of other noninvasive imaging methods measuring various aspects of tumour vasculature, metabolism and proliferation. This is because a drug can induce many different responses in the tumour that are specific but also nonspecific to the mechanism of action. Therefore, a more generic marker of cell viability or vascular status is likely to be the better universal early-response marker

Table 4 Effect of different compound classes on minimally invasive parameters 2–3 days after treatment initiation

Compound	Molecular target	iAUEC	Ktrans	BVol	IFP	FDG	FLT
EPO906	Micro-tubules	1.03	0.82	0.7*	0.41*	0.54*	0.78*
PTK787	VEGF-R	0.80*	0.37*	0.97	0.64*	0.94	–
RAD001	mTOR	1.13	1.34	1.1	0.74	0.70* HS 1.02 LS	0.82*

The values shown reflect the average T/Cs (fractional change in parameter for compound-treated divided by vehicle-treated) determined in several different experiments across one or frequently two different tumour models (BN472, B16/BL6, RIF-1, H-596, HCT-116, KB-31), where *indicates there was a significant difference relative to vehicle. For RAD001 on FDG, two values are shown to reflect the very different behaviour in low- and high-sensitivity models

of successful therapy. Unfortunately, IFP measurements in the clinic are not always feasible because of the position of the tumour, patient agreement, or indeed the perception of the treating clinician. This is understandable, and it means we still await the development of a fast, safe, easy, noninvasive and robust method to measure a parameter that always changes in the same direction when there is a significant response of the tumour to therapy. Therefore, for now, the optimal early-response biomarker for a particular compound remains to be identified preclinically before clinical trials are in place, because, as Table 4 illustrates, a biomarker change is not always predictable.

References

Ferretti S, Allegrini PR, O'Reilly T, Schnell C, Stumm M, Wartmann M, Wood J and McSheehy PM (2005) Patupilone induced vascular disruption in orthotopic rodent tumor models detected by magnetic resonance imaging and interstitial fluid pressure. Clin Cancer Res 11:7773–7784

Jain RK (1994) Barriers to drug deliver in solid tumors. Sci Am 271:58–65

Minimally Invasive Biomarkers for Therapy Monitoring

Lee L, Sharma S, Morgan B, Allegrini P, Schnell C, Brueggen J, Cozens R, Horsfield M, Guenther C, Steward WP, Drevs J, Lebwohl D, Wood J, McSheehy PM (2006) Biomarkers for assessment of pharmacologic activity for a vascular endothelial growth factor (VEGF) receptor inhibitor, PTK787/ZK 222584 (PTK/ZK): translation of biological activity in a mouse melanoma metastasis model to phase I studies in patients with advanced colorectal cancer with liver metastases. Cancer Chemother Pharmacol 57:761–71

Mankoff DA, Muzi M, Krohn KA (2003) Quantitative positron emission tomography imaging to measure tumor response to therapy: what is the best method? Mol Imaging Biol 5:281–285

McSheehy P, Allegrini P, Honer M, Ebenhan T, Ametaby S, Schubiger P, Schnell C, Stumm M, O'Reilly T, Lane H (2007) Monitoring the activity of the mTor pathway inhibitor RAD001 (everolimus) non-invasively by functional imaging. Targeted Oncol 2:130–131

Milosevic M, Fyles A, Hedley D, Pintilie M, Levin W, Manchul L, Hill R (2001) Interstitial fluid pressure predicts survival in patients with cervix cancer independent of clinical prognostic factors and tumor oxygen measurements. Cancer Res 61:6400–6405

Rudin M, McSheehy PM, Allegrini PR, Rausch M, Baumann D, Becquet M, Brecht K, Brueggen J, Ferretti S, Schaeffer F, Schnell C, Wood J (2005) PTK787 / ZK222584, a tyrosine kinase inhibitor of vascular endothelial growth factor receptor, reduces uptake of the contrast agent GdDOTA by murine orthotopic B16/BL6 melanoma tumors and inhibits their growth in vivo. NMR Biomed 18:308–321

Rutz HP (1999) A biophysical basis of enhanced interstitial fluid pressure in tumors. Med Hypotheses 53:526–529

Schnell CR, Stauffer F, Allegrini PR, O'Reilly T, McSheeny PMJ, Dartois C, Stumm M, Cozens R, Littlewood-Evans A, García-Echeverría C, Maira S-M (2008) Effects on the dual pan-class I PI3K/mTor inhibitor NVP-BEZ235 on the tumour vasculature: implications for clinical imaging. Cancer Res 68 (in press)

Taghian AG, Abi-Raad R, Assaad SI, Casty A, Ancukiewicz M, Yeh E, Molokhia P, Attia K, Sullivan T, Kuter I, Boucher Y, Powell SN (2005) Paclitaxel decreases the interstitial fluid pressure and improves oxygenation in breast cancers in patients treated with neoadjuvant chemotherapy: clinical implications. J Clin Oncol 23:1951–1961

Troprès I, Grimault S, Vaeth A, Grillon E, Julien C, Payen JF, Lamalle L, Décorps M (2001) Vessel size imaging. Magn Reson Med 45:397–408

Willett CG, Boucher Y, di Tomaso E, Duda DG, Munn LL, Tong RT, Chung DC, Sahani DV, Kalva SP, Kozin SV, Mino M, Cohen KS, Scadden DT, Hartford AC, Fischman AJ, Clark JW, Ryan DP, Zhu AX, Blaszkowsky LS, Chen HX, Shellito PC, Lauwers GY, Jain RK (2004) Direct evidence that the VEGF-specific antibody bevacizumab has antivascular effects in human rectal cancer. Nat Med 10:145–147

Ernst Schering Foundation Symposium Proceedings, Vol. 4, pp. 189–203
DOI 10.1007/2789_2008_094
© Springer-Verlag Berlin Heidelberg
Published Online: 06 June 2008

Use of Metabolic Pathway Flux Information in Anticancer Drug Design

L.G. Boros[✉], T.F. Boros

Harbour-UCLA Medical Center RB1, Los Angeles Biomedical Research Institute,
1124 West Carson Street, 90502 Torrance, USA
email: *lboros@sidmap.com*

This article is to commemorate Dr. Boros, Ferenc János, a dedicated physician to his patients, a loving father and a memorable brother to the authors. Dr. Boros passed away of cancer in 2006 and this article is dedicated to his memory and to those who have suffered from this illness.

1	Introduction	190
2	Metabolic Profiles of Tumor Cells	190
3	Tumor Growth-Promoting Signals Influence Phenotype Expression via Nonoxidative Metabolism	193
4	Tumor Growth-Inhibiting Signals Limit Nonoxidative Metabolism	195
5	Metabolic Constraints of Cell Growth	195
6	Concluding Remarks	199
References		200

Abstract. The metabolic phenotype of tumor cells promote the proliferative state, which indicates that (a) cell transformation is associated with the activation of specific metabolic substrate channels toward nucleic acid synthesis and (b) increased expression phosphorylation, allosteric or transcriptional regulation of intermediary metabolic enzymes and their substrate availability together mediate unlimited growth. It is evident that cell transformation due to various K-*ras* point mutations is associated with the activation of specific metabolic substrate channels that increase glucose channeling toward nucleic acid synthesis. Therefore, phosphorylation, allosteric and transcriptional regulation of intermediary metabolic enzymes and their substrate availability together mediate cell transformation and growth. In this review, we summarize opposite changes in metabolic phenotypes induced by various cell-transforming agents, and tumor growth-inhibiting drugs or phytochemicals, or novel synthetic antileukemic drugs such as imatinib mesylate (Gleevec). Metabolic enzymes that further incite growth signaling pathways and thus promote malignant cell transformation serve as high-efficacy nongenetic novel targets for cancer therapies.

1 Introduction

There are over 300 tumor-inducing genes, environmental factors, and signal transduction pathways identified to date. The most common genetic abnormalities reported include frequent mutations of the K-*ras*, *p53*, *p16*, and *Smad4* genes, which were reported to be associated with accelerated cancer progression and poor prognosis (Yatsuoka et al. 2000; Sakai et al. 2000). Genetic abnormalities that influence cellular responses to hormonal growth regulators and their signaling pathways have been reported in connection with all major tumor types (Szabo et al. 2000; Ozen and Pathak 2000; Largaespada 2000; Martin and Weber 2000; Jung and Messing 2000; Issa 2000). The general understanding of malignant tumors emphasizes strong genetic regulation of human cell functions. In this chapter, we emphasize that the diverse mechanisms of tumor induction commonly influence metabolism and thus the potential for differentiation, cell cycle arrest, and apoptosis (Fig. 1).

In spite of their great genetic potential to express different phenotypes corresponding to the different degree of differentiation, a great majority of human tumors exhibit a simple high glucose-utilizing phenotype and poor differentiation (Warburg 1930, 1956; Krebs et al. 1950). Tumor cells primarily utilize glucose for intracellular macromolecule synthesis, mainly nucleotide synthesis, as well as other anabolic reactions such as lipid and protein synthesis (Horecker 1965).

2 Metabolic Profiles of Tumor Cells

Malignant cells compete for glucose ten to 50 times more intensely than the surrounding normal tissue (Warburg 1930, 1956; Krebs et al.

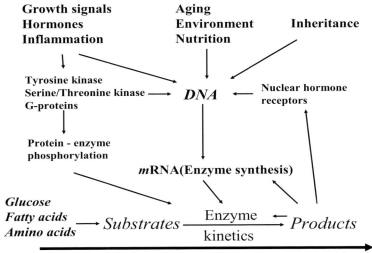

Fig. 1. Cell growth-controlling signals regulate metabolic enzymes through protein phosphorylation or via the transcription of genes. Nutrients affect cell metabolism by altering substrate availability for macromolecule synthesis and energy production. Glucose and nutrient components directly influence metabolism as allosteric regulators of enzymes of intermediary metabolism. Substrate flow is also influenced by substrate/product ratios and by intracellular substrates and products acting on nuclear receptors. The overall metabolic effect of these regulatory pathways can be determined by tracer substrate-based metabolomics using mass spectrometry or NMR by measuring the flux of nutrient carbons through various metabolic pathways as the flow of carbons toward the synthesis of different macromolecules depicts the metabolic phenotype of cancer cells

1950). This is the biochemical basis of tumor diagnosis with the radioactive glucose analog tracer ^{18}fluoro-deoxy-glucose using positron emission tomography (FDG-PET). An increased rate of glucose accumulation strongly correlates with poor biological behavior and invasiveness (Raylman et al. 1995; Torizuka et al. 1995; Strauss and Conti 1991; Bares et al. 1994). Proliferating cells are highly dependent on purine and pyrimidine synthesis as well as ribose for nucleic acids. Glucose is a particularly important substrate from which tumor cells produce ribose, the backbone of RNA and DNA molecules.

The advantage of stable isotope-based mass spectrometry is in its ability to determine the quantity and position of label incorporation into macromolecules that promote cell cycle progression and proliferation in tumor cells. Such capability was first demonstrated in the study of the glycogen synthesis pathway using uniformly labeled U ^{13}C-glucose in experimental animals (Katz et al. 1989). In consequent studies, the analysis of label incorporation into glutamate for the purpose of sampling α-keto-glutarate pools provided the basis for the study of pyruvate dehydrogenase (PDH) and pyruvate carboxylase activity as well as the anaplerotic flux of the TCA cycle (Lee 1993). Taking advantage of the knowledge of mass isotopomer distribution of a biomolecule, fatty acid synthesis can directly be studied using ^{13}C-labeled acetate or deuterated water (D_2O) (Lee et al. 1995). [1, 2-$^{13}C_2$]-D-glucose is also commonly used in metabolic phenotyping studies, which provides unique information on the carbon flow through the oxidative and nonoxidative branches of the pentose cycle by its labeling patterns in various intermediates and products required for cancer growth. For example, [1, 2-$^{13}C_2$]-D-glucose readily labels oxaloacetate when it is carboxylated via pyruvate carboxylase and carbon 2 and 3 of α-keto-glutarate within the TCA cycle. On the other hand [1, 2-$^{13}C_2$]-D-glucose is converted to acetyl-CoA by pyruvate dehydrogenase, which readily labels carbons 4 and 5 of α-keto-glutarate and also generates ^{13}C-labeled fatty acids depicting lipid synthesis as well as acetate enrichment of fatty acids from glucose (Boros et al. 2001a).

Alternate substrates using labeled fatty acids, glycerol, or ribose can be developed to further characterize the role of these nutrients in different metabolic phenotypes. Such studies can reveal metabolic activity and the identification of structural and regulatory macromolecules that are synthesized using primarily glucose carbons in human cells (Horecker et al. 1958; Katz and Rognstad 1967). Extensive reviews and methodologic reports published in specialty journals have demonstrated the usefulness of the stable isotope method in combination with biologic mass spectrometry in mammalian cell metabolic studies to reveal glucose-derived specific synthesis pathways of nucleotides, lipids,

Metabolic Pathway Flux Information in Cancer

and amino acids associated with tumor proliferation, differentiation, or apoptosis (Lee et al. 1992, 1995; Kingsley-Hickman et al. 1990).

In general, molecular biology methods using PCR hybridization and sequencing of DNA and RNA reveal genetic abnormalities and depict the expression of genes or identifies gene sequences being expressed. Proteomics methods using chromatography in combination with mass spectrometry reveal peptide, protein composition and structure of cell skeletal and enzyme proteins. Enzymology and enzyme activity measurements readily determine enzyme activities and reveal metabolic control characteristics of enzymes on substrate flow. On the other hand, tracer substrate-based biological mass spectrometry by the intracellular labeling of RNA, DNA, amino acids, lipids as intermediates of tracer-labeled substrate molecules determine substrate flux through metabolic pathways and thus identify the contribution of substrates to macro-molecule synthesis during genetic and signaling events in fully functioning cells. Therefore, the stable isotope tracer substrate approach in combination with biologic mass spectrometry complements molecular genetic studies.

In summary, although molecular genetic studies provide information on the regulatory action of genes, hormones, and signal transduction pathways, it is the nutrients and substrate availability that ultimately influence the rate and patterns of substrate distribution and tumor cell functions and proliferation. The most measurable significance of directly characterizing substrate distribution in tumor cells is the ability to determine the overall metabolic effect of tumor growth-regulating agents and substrate availability, which constitute together the most fundamental constraints of growth in all living organisms.

3   Tumor Growth-Promoting Signals Influence Phenotype Expression via Nonoxidative Metabolism

There are two major growth signal types: one, such as steroid hormones, which acts on intracellular receptors and influences gene expression, and the other, such as transforming growth factors, which acts on cell surface receptors and influences multiple enzyme activities by protein phosphorylation. For example, transforming growth factor (TGF-β) on

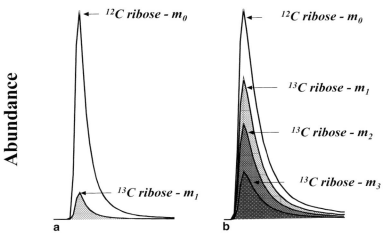

Fig. 2a,b. Ribose mass spectra of TGF-β-treated H441 lung epithelial carcinoma cells in culture. The chemical ionization mass spectral analysis reveals a significant increase in ^{13}C-carbon deposition into RNA from glucose after TGF-β treatment as shown by the increase in the shaded areas on the spectrum (**b**). Ribose mass spectra in nature show less ^{13}C abundance (**a**). Ribose mass spectra using [1,2-^{13}C$_2$]-D-glucose tracer (50%) and biologic mass spectrometry; and mass isotopomers shown as m_0, m_1, m_2, and m_3 represent ribose molecules in RNA with 0, 1, 2, or 3 ^{13}C substitutions, respectively

lung epithelial carcinoma cells uses the tyrosine kinase signaling mechanism through the cell surface TGF-β receptor family and promotes the invasive transformation of various human cells (Hojo et al. 1999). The accumulation of glucose carbons in nucleic acid ribose demonstrated a significant, dose-dependent increase in response to this growth factor (Fig. 2).

Concomitant metabolic changes in response to TGF-β treatment included decreased complete glucose oxidation in the TCA cycle, indicating that invasive cell transformation is accompanied by nonoxidative metabolic changes and increased glucose utilization toward anabolic synthetic reactions of nucleotides (Boros et al. 2000c). This increase in the nonoxidative metabolism of glucose in the pentose cycle provided an explanation at the molecular level for the principal metabolic

Metabolic Pathway Flux Information in Cancer 195

disturbance observed in human tumors: increased glucose uptake with increased glucose utilization for nucleic acid synthesis and decreased glucose oxidation. These metabolic changes also reflect how tumor cells are capable of dividing rapidly in the hypoxic environment.

4 Tumor Growth-Inhibiting Signals Limit Nonoxidative Metabolism

Genistein, the natural tumor growth-regulating agent found in soy bean, has marked tyrosine kinase- and protein kinase-inhibiting properties (El-Zarruk and van den Berg 1999; Waltron and Rozengurt 2000), resulting in cell cycle arrest (Lian et al. 1998) and limiting angiogenesis (Zhou et al. 1999) in several tumor models. Genistein decreases glucose uptake and glucose carbon incorporation into nucleic acid ribose adenocarcinoma cells (Fig. 3) (Boros et al. 2001a; Katz and Rognstad 1967).

The opposite changes in glucose carbon deposition into nucleic acid ribose, lactate, glutamate, and fatty acids after treating tumor cells with growth-promoting (Boros et al. 2000c) or growth-inhibiting agents (Boros et al. 2001a) indicate that glucose carbon redistribution between major metabolic pathways plays a critical role in tumor cell proliferation.

5 Metabolic Constraints of Cell Growth

There is specific distribution of substrates between macromolecule synthesis pathways and energy production driven by the metabolic needs of the major cellular phenotypes. Substrate availability directly determines the rate of proliferation and differentiation, which are characterized as metabolic constraints for the production of energy and substrates necessary for cells to function under different pathologic conditions. It is clear from work on pancreatic, lung, and other epithelial malignancies that invasive transformation is associated with characteristic metabolic changes. The typical metabolic phenotypes related to tumor cell formation or death are summarized as follows:

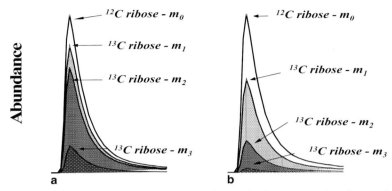

Fig. 3a,b. Ribose mass spectral changes after genistein treatment in cultures of MIA pancreatic adenocarcinoma cells. The chemical ionization mass spectral analysis reveals a significant decrease in ^{13}C-carbon deposition into RNA from glucose after genistein treatment (**b**) in reference to MIA cells grown in fetal bovine serum containing media (**a**). Ribose mass spectra using [1,2-^{13}C$_2$]-D-glucose tracer (50%) and biologic mass spectrometry; mass isotopomers shown as m_0, m_1, m_2, and m_3 represent ribose molecules in RNA with 0, 1, 2, or 3 ^{13}C substitutions, respectively

1. Proliferation and S phase cycle dominance requires the continuous flow of glucose carbons through the oxidative and nonoxidative steps of the pentose cycle for de novo DNA and RNA synthesis. Therefore metabolic constraints exist in the pentose cycle, as demonstrated in H441 lung adenocarcinoma, K562 myeloid blasts after TGF-β or isofenphos pesticide treatments via tyrosine/protein kinase activation (Boros et al. 2000c).
2. Differentiation G0–G1 phase cycle dominance requires a shift of glucose carbons to direct oxidation through glucose-6P dehydrogenase (G6PDH) and recycling of ribose carbons back into glycolysis; increased lipid and amino acid synthesis from glucose. Therefore, metabolic constraints exist in the oxidative pentose cycle, direct glucose oxidation and glycolysis, as well as pyruvate kinase. Genistein, Avemar and STI571 typically inhibit various tyrosine/protein kinases (Boros et al. 2001a; Rais et al. 1999) along these pathways.

Metabolic Pathway Flux Information in Cancer   197

3. Cell cycle arrest and G0 or G2-M phase cycle dominance are characterized by limited carbon flow through both the oxidative and the nonoxidative branches of the pentose cycle. Limited RNA/DNA synthesis and limited NADPH production restrain nonoxidative metabolism, transketolase and the oxidative branch of the pentose cycle. Cycle arrest may occur from the simple results of enzyme inhibition in MIA pancreatic carcinoma and Ehrlich's ascites cells in mice after oxythiamine treatment combined with DHEA-S while no signal transducer pathways are needed (Chesney et al. 1999; Boros et al. 2000a).
4. Apoptosis or G0 cycle arrest is induced by limited glucose availability or the direct inhibition of glycolytic, pentose and TCA cycle enzymes (Shim et al. 1998; Osthus et al. 2000) using false glucose analogs such as deoxy-D-glucose (DOG).

Macromolecule synthesis is highly dependent on the availability of glucose carbons and increased activity of glycolytic, pentose, and TCA cycle enzymes. As a feedback mechanism, the production of intermediary metabolites also regulates gene expression through intracellular nuclear receptors. In support of our model, there are strong interactions between newly discovered signal transduction pathways and fundamental metabolic pathways such as glycolysis and the pentose cycle. Glucose deprivation is capable of inducing apoptosis of tumor cells on its own, even when other nutrients are plentiful (Spitz et al. 2000; Shim et al. 1998) and exerts as the strongest metabolic constraint of cellular growth. For example, the c-*myc* oncogene directly regulates glucose transporter 1 and glycolytic gene expression in several tumor cells (Osthus et al. 2000). Our model emphasizes that there is an apparent functional hierarchy within growth signaling, the translation of genes into proteins and enzymes that regulate various key metabolic pathways for macromolecule synthesis and energy production. The flow of information from the exterior of cells using specific signal transducer pathways and the flow of substrates to restrain these signals are key elements in the regulation of cell function. These events also regulate the optimum activity of metabolic enzymes, which also represent key regulatory check posts of carbon redistribution between major metabolic pathways. The direct and indirect interactions between signal transducer pathways, substrates, and their intracellular target enzymes allow a fine regulatory mechanism in a functional hierarchy of genes and proteins, which together control cellular events. Accordingly, multiple genetic alterations and signaling pathways that cause tumor development directly affect glycolysis (Kun-Schughart et al. 2000), the cellular response to hypoxia (Boros et al. 2000c), and the ability of tumor cells to recruit new blood vessels (Oku et al. 1998).

Examples of metabolic constraints via enzyme activation and substrate availability of cell transformation and growth are overwhelming in the medical literature. For example, the E7 oncoprotein, encoded by the oncogenic human papillomavirus (HPV) type 16, binds to the glycolytic enzyme type M_2 pyruvate kinase (M_2-PK). Pyruvate kinase exhibits a tetrameric form with a high affinity to its substrate phosphoenolpyruvate in normal cells and a dimeric form with a low affinity to phosphoenolpyruvate in tumor cells. As a result, tumor cells accumulate high levels of phosphorylated glycolytic metabolites to support nucleic acid synthesis at the expense of the carbon pools for lipid and amino acid syntheses. Investigations of HPV-16 E7 mutants and the nononcogenic HPV-11 subtype suggest that the interaction of HPV-16 E7 with M_2-PK is linked to the transforming potential of the viral oncoprotein through metabolic adaptive changes (Zwerschke et al. 1999). Another example involves the *c-Myc* oncogenic transcription factor, which regulates lactate dehydrogenase A and induces lactate overproduction. c-Myc, however, also controls other genes regulating glucose metabolism. In Rat1a fibroblasts and murine livers overexpressing *c-Myc*, the mRNA levels of the glucose transporter GLUT1, phosphoglucose isomerase, phosphofructokinase, glyceraldehyde-3-phosphate dehydrogenase, phosphoglycerate kinase, and enolase were elevated. *c-Myc* directly transactivated genes encoding GLUT1, phosphofructokinase, and enolase and eventually increased glucose uptake (Osthus et al. 2000). These metabolites are necessary for nonoxidative nucleic acid synthesis (Chesney et al. 1999), which can be precisely characterized using stable isotopes and GC/MS. Such an approach is an excellent screening method for studying the direct metabolic effects of new antitumor drugs.

One specific target site where the metabolic constraint approach could bring new remedies for the treatment of cancer has been established within the nonoxidative reactions of the pentose cycle. Transketolase has been identified as the key enzyme in the regulation of glucose carbon recruitment for the de novo synthesis of nucleic acid ribose (Boros et al. 2000a), and it also has an exceptionally high growth control coefficient in in vivo tumor proliferation. Because the blood plasma of mammalians contains only a very limited supply of five carbon sugars, it is inevitable that glucose recruitment for nucleic acid synthesis is one of the key metabolic regulatory steps where effective tumor growth control can be achieved. In previous studies, we successfully applied the chemically modified transketolase co-factor, oxythiamine, for the treatment of experimental cancer in animals (Boros et al. 1997). Oxythiamine treatment induced a dose-dependent arrest in the progression of the cell cycle in Ehrlich's tumor-hosting animals (Rais et al. 1999). Recent studies using novel tumor growth-inhibiting agents, such as the wheat germ extract Avemar for the treatment of human colorectal malignancies with advanced liver metastases (Boros et al. 2000b, 2001b; Jakab et al. 2000) or STI571 (Glivec) for the treatment of chronic myeloid leukemias (Boren et al. 2001), reveal strong inhibitory action on glucose use for nucleic acid synthesis as the central mechanism of antiproliferative action.

6   Concluding Remarks

Tumor cells assume their unique characteristics according to their diverse genetic aberrations. Their invasive and proliferative characteristics, however, are limited by the availability of substrates, nutrients, and metabolic pathway enzyme activities. Based on these factors, tumor cells exhibit distinct metabolic phenotypes determining the rate of proliferation, apoptosis, cell cycle arrest, and differentiation. Hormones, signaling pathways, environmental factors, and nutritional habits have a strong influence on these metabolic phenotypes. Understanding the adaptive metabolic changes in glycolysis and anabolic reactions in response to tumor growth-modulating agents is fundamental to the understanding tumor pathophysiology. The proposed metabolic constraint model of tumor growth and death permits a wide range of basic and clinical studies in order to develop new strategies to revert tumor-specific metabolic changes for the benefit of the human host.

Acknowledgements. This work was, in part, supported by the PHS M01-RR00425 of the General Clinical Research Unit, by NIH-AT00151, by P01-CA42710 of the UCLA Clinical Nutrition Research Unit Stable Isotope Core, its 009826-00-00 Preliminary Feasibility grant to LGB and by P01 AT003960-01 UCLA Center for Excellence in Pancreatic Diseases, Metabolomics Core, and a grant from the Hirshberg Foundation for Pancreatic Cancer Research.

References

Bares R, Klever P, Hauptmann S et al (1994) F-18 fluorodeoxyglucose PET in vivo evaluation of pancreatic glucose metabolism for detection of pancreatic cancer. Radiology 192:79–86

Boren J, Cascante M, Marin S et al (2001) Gleevec (STI571) influences metabolic enzyme activities and glucose carbon flow toward nucleic acid and fatty acid synthesis in myeloid tumor cells. J Biol Chem 276:37747–37753

Boros LG, Puigjaner J, Cascante M et al (1997) Oxythiamine and dehydroepiandrosterone inhibit the nonoxidative synthesis of ribose and tumor cell proliferation. Cancer Res 57:4242–4248

Boros LG, Comin B, Boren J et al (2000a) Overexpression of transketolase: a mechanism by which thiamine supplementation promotes cancer growth (abstract). Proc Am Assoc Cancer Res 41:666

Boros LG, Lee W-NP, Hidvegi M et al (2000b) Metabolic effects of fermented wheat germ extract with anti-tumor properties in cultured MIA pancreatic adenocarcinoma cells. Pancreas 21:433

Boros LG, Torday JS, Lim S et al (2000c) Transforming growth factor beta2 promotes glucose carbon incorporation into nucleic acid ribose through the nonoxidative pentose cycle in lung epithelial carcinoma cells. Cancer Res 60:1183–1185

Boros LG, Bassilian S, Lim S et al (2001a) Genistein inhibits nonoxidative ribose synthesis in MIA pancreatic adenocarcinoma cells: a new mechanism of controlling tumor growth. Pancreas 22:1–7

Boros LG, Lapis K, Szende B et al (2001b) Wheat germ extract decreases glucose uptake and RNA ribose formation but increases fatty acid synthesis in MIA pancreatic adenocarcinoma cells. Pancreas 23:141–147

Metabolic Pathway Flux Information in Cancer

Chesney J, Mitchell R, Benigni F et al (1999) An inducible gene product for 6-phosphofructo-2-kinase with an AU-rich instability element: role in tumor cell glycolysis and the Warburg effect. Proc Natl Acad Sci U S A 96:3047–3052

El-Zarruk AA, van den Berg HW (1999) The anti-proliferative effects of tyrosine kinase inhibitors towards tamoxifen-sensitive and tamoxifen-resistant human breast cancer cell lines in relation to the expression of epidermal growth factor receptors (EGF-R) and the inhibition of EGF-R tyrosine kinase. Cancer Lett 142:185–193

Hojo M, Morimoto T, Maluccio M et al (1999) Cyclosporine induces cancer progression by a cell-autonomous mechanism. Nature 397:530–534

Horecker BL (1965) Pathways of carbohydrate metabolism and their physiological significance. J Chem Educ 42:244–253

Horecker BL, Domagk G, Hiatt HH (1958) A comparison of C14-labeling patterns in deoxyribose and ribose in mammalian cells. Arch Biochem Biophys 78:510–517

Issa JP (2000) The epigenetics of colorectal cancer. Ann N Y Acad Sci 910:140–153

Jakab F, Mayer A, Hoffmann A et al (2000) First clinical data of a natural immunomodulator in colorectal cancer. Hepatogastroenterology 47:393–395

Jung I, Messing E (2000) Molecular mechanisms and pathways in bladder cancer development and progression. Cancer Control 7:325–334

Katz J, Rognstad R (1967) The labeling of pentose phosphate from glucose-^{14}C and estimation of the rates of transaldolase, transketolase, the contribution of the pentose cycle and ribose phosphate synthesis. Biochemistry 6:2227–2247

Katz J, Lee W-NP, Wals PA et al (1989) Studies of glycogen synthesis and the Krebs cycle by mass isotopomer analysis with U-^{13}C-glucose in rats. J Biol Chem 264:12994–13001

Kingsley-Hickman PB, Ross B, Krick T (1990) Hexose monophosphate shunt measurement in cultured cells with [1-^{13}C]glucose: correction for endogenous carbon sources using [6-^{13}C]Glucose. Anal Biochem 185:235–237

Krebs ET Jr, Krebs ET Sr, Beard HH (1950) The unitarian or trophoblastic thesis of cancer. Med Rec 163:150–171

Kunz-Schughart LA, Doetsch J, Mueller-Klieser W et al (2000) Proliferative activity and tumorigenic conversion: impact on cellular metabolism in 3-dimensional culture. Am J Physiol Cell Physiol 278:C765–C780

Largaespada DA (2000) Genetic heterogeneity in acute myeloid leukemia: maximizing information flow from MuLV mutagenesis studies. Leukemia 14:1174–1184

Lee WN, Bergner EA, Guo ZK (1992) Mass isotopomer pattern and precursor–product relationship. Biol Mass Spectrom 21:114–122

Lee W-NP (1993) Analysis of tricarboxylic acid cycle using mass isotopomer ratios. J Biol Chem 268:25522–25526

Lee W-NP, Byerley LO, Bassilian S et al (1995) Isotopomer study of lipogenesis in human hepatoma cells in culture: contribution of carbon and reducing hydrogen from glucose. Anal Biochem 226:100–112

Lian F, Bhuiyan M, Li YW et al (1998) Genistein-induced G2-M arrest, p21WAF1 upregulation and apoptosis in a non-small-cell lung cancer cell line. Nutr Cancer 31:184–191

Martin AM, Weber BL (2000) Genetic and hormonal risk factors in breast cancer. J Natl Cancer Inst 92:1126–1135

Oku T, Tjuvajev JG, Miyagawa T et al (1998) Tumor growth modulation by sense and antisense vascular endothelial growth factor gene expression: effects on angiogenesis, vascular permeability, blood volume, blood flow, fluorodeoxyglucose uptake and proliferation of human melanoma intracerebral xenografts. Cancer Res 58:4185–4192

Osthus RC, Shim H, Kim S et al (2000) Deregulation of glucose transporter 1 and glycolytic gene expression by c-Myc. J Biol Chem 275:21797–21800

Ozen M, Pathak S (2000) Genetic alterations in human prostate cancer: a review of current literature. Anticancer Res 20:1905–1912

Rais B, Comin B, Puigjaner J et al (1999) Oxythiamine and dehydroepiandrosterone induce a G1 phase cycle arrest in Ehrlich's tumor cells through inhibition of the pentose cycle. FEBS Lett 456:113–118

Raylman RR, Fisher SJ, Brown RS et al (1995) Fluorine-18-fluorodeoxyglucose-guided breast cancer surgery with a positron-sensitive probe: validation in preclinical studies. J Nucl Med 36:1869–1874

Sakai Y, Yanagisawa A, Shimada M et al (2000) K-ras gene mutations and loss of heterozygosity at the p53 gene locus relative to histological characteristics of mucin-producing tumors of the pancreas. Hum Pathol 31:795–803

Shim H, Chun YS, Lewis BC et al (1998) A unique glucose-dependent apoptotic pathway induced by c-Myc. Proc Natl Acad Sci U S A 95:1511–1516

Spitz DR, Sim JE, Ridnour LA et al (2000) Glucose deprivation-induced oxidative stress in human tumor cells. A fundamental defect in metabolism? Ann N Y Acad Sci 899:349–362

Strauss LG, Conti PS (1991) The application of PET in clinical oncology. J Nucl Med 32:623–648

Szabo C, Masiello A, Ryan JF et al (2000) The Breast Cancer Information Core: database design, structure and scope. Hum Mutat 16:123–131

Metabolic Pathway Flux Information in Cancer

Torizuka T, Tamaki N, Inokuma T et al (1995) In vivo assessment of glucose metabolism in hepatocellular carcinoma with FDG-PET J Nucl Med 36:1811–1817

Waltron RT, Rozengurt E (2000) Oxidative stress induces protein kinase D activation in intact cell: involvement of Src and dependence on protein kinase C. J Biol Chem 275:17114–17121

Warburg O (1930) The metabolism of tumors. London, Costable

Warburg O (1956) On the origin of cancer cells. Science 123:309–314

Yatsuoka T, Sunamura M, Furukawa T et al (2000) Association of poor prognosis with loss of 12q, 17p and 18q and concordant loss of 6q/17p and 12q/18q in human pancreatic ductal adenocarcinoma. Am J Gastroenterol 95:2080–2085

Zhou JR, Gugger ET, Tanaka T et al (1999) Soybean phytochemicals inhibit the growth of transplantable human prostate carcinoma and tumor angiogenesis in mice. J Nutr 129:1628–1635

Zwerschke W, Mazurek S, Massimi P et al (1999) Modulation of type M2 pyruvate kinase activity by the human papillomavirus type 16 E7 oncoprotein. Proc Natl Acad Sci U S A 96:1291–1296

Ernst Schering Foundation Symposium Proceedings, Vol. 4, pp. 205–226
DOI 10.1007/2789_2008_095
© Springer-Verlag Berlin Heidelberg
Published Online: 03 July 2008

Cancer Diagnostics Using ^{1}H-NMR-Based Metabonomics

K. Odunsi[✉]

Departments of Gynecologic and Oncology Immunology, Roswell Park Cancer Institute,
Elm and Carlton Streets, 14261 Buffalo, USA
email: *Kunle.Odunsi@Roswellpark.org*

1	Background	206
2	Current Status of Early Detection of EOC in the General Population	207
3	NMR-Based Metabonomics for the Analysis of Biofluids	208
4	^{1}H-NMR Analysis of Plasma and Cancer Detection	209
5	^{1}H-NMR-Based Metabonomics for Ovarian Cancer Detection	210
5.1	^{1}H-NMR Spectroscopic Analysis of the Serum Samples	211
5.2	Data Reduction of NMR Data	212
5.3	Pattern Recognition Analysis of the ^{1}H-NMR Spectra	212
5.4	Soft Independent Modeling of Class Analogy	215
5.5	Receiver Operating Characteristic Curve Analysis	218
5.6	Analysis of Spectral Pattern Differences	218
5.7	Validation of EOC Results in an Independent Set of Serum Specimens	220
5.8	Mass Spectrometry-Based Metabolic Profiling in Ovarian Cancer	220
6	Conclusions and Future Directions	222
References		223

Abstract. For several solid human malignancies, currently available serum biomarkers are insufficiently reliable to distinguish patients from healthy individuals. Metabonomics, the study of metabolic processes in biologic systems, is based on the use of ^{1}H-NMR spectroscopy and multivariate statistics for biochemical data generation and interpretation and may provide a characteristic fingerprint in disease. Here we review our initial experiences utilizing the metabonomic approach for discriminating sera from women with epithelial ovarian cancer (EOC) from healthy controls. ^{1}H-NMR spectroscopic analysis was performed on preoperative serum specimens of 38 EOC patients, 12 patients with benign ovarian cysts and 53 healthy women. PCA analysis allowed correct separation of all serum specimens from 38 patients with EOC (100%) from all of the 21 premenopausal normal samples (100%) and from all the sera from patients with benign ovarian disease (100%). In addition, it was possible to correctly separate 37 of 38 (97.4%) cancer specimens from 31 of 32 (97%) postmenopausal control sera. ROC analysis indicated that the sera from patients with and without disease could be identified with 100% sensitivity and specificity at the ^{1}H-NMR regions 2.77 parts per million (ppm) and 2.04 ppm from the origin (AUC of ROC curve = 1.0). These findings indicate that the ^{1}H-NMR metabonomic approach deserves further evaluation as a potential novel strategy for the early detection of EOC.

1 Background

Cancer is a major public health problem. Current estimates suggest that approximately three out of every ten individuals will be diagnosed with cancer at some point during their lifetime (Wingo et al. 1995). Screening and early detection are two strategies with the potential to reduce morbidity and mortality from a particular cancer among the screened population. Although significant advances have been made in screening/early detection of cancers of the breast, cervix, skin, and colon, there are no reliable early detection strategies for cancers of the pancreas, lungs and epithelial ovarian cancer (EOC). This review focuses on the use of ^{1}H-NMR-based metabonomics as a potential method for early detection of EOC.

Ovarian cancer is the leading cause of death from gynecologic malignancies. There are more than 23,000 cases annually in the United States, and approximately 14,000 women can be expected to die from the disease in 2007. Survival rates remain disappointing for patients with advanced EOC and primary peritoneal carcinomas despite modest improvements in response rates, progression-free survival and median survival using adjuvant platinum and paclitaxel chemotherapy following cytoreductive surgery (Armstrong et al. 2006; McGuire et al. 1996). This has been attributed to two important reasons. First, in contrast to most other solid tumors, more than 75% of EOC patients present with advanced-stage disease (FIGO III or IV). Whereas the small proportion of patients with accurately diagnosed stage I disease have 5-year survival rates in excess of 90% (Young et al. 1990), the survival rate for women diagnosed with distant disease is only 25%. Secondly, although most patients with advanced disease initially respond to platinum- and paclitaxel-based chemotherapy including complete responses, the relapse rate is approximately 85% (Greenlee et al. 2001). Within 2 years of cytoreductive surgery and systemic chemotherapy, tumors usually recur and once relapse occurs, there is no known curative therapy. The link between stage and mortality suggests that early detection may have a significant impact on disease morbidity and mortality in EOC. The need for early detection is especially acute in women who have a high risk of ovarian cancer due to family or personal history of cancer, and for women with a genetic predisposition to cancer due to abnormalities in predisposition genes such as *BRCA1* and *BRCA2*.

Although a number of potential early detection strategies have been studied in EOC (Menon and Jacobs 2000), these have shown only limited promise. The ideal test for the early detection of EOC should be noninvasive, acceptable to the screened population, have high validity, and have a relatively low cost. The application of novel approaches such as functional genomics, proteomics and metabonomics may improve the ability to detect EOC at an early stage, with the potential of reducing morbidity and mortality from the disease.

2 Current Status of Early Detection of EOC in the General Population

The majority of patients with EOC come from low-risk families and are usually diagnosed due to symptoms of advanced disease. Current candidate strategies for early detection of EOC in this population are based on biochemical tumor markers evaluated mainly in the blood and biophysical markers assessed by ultrasound and/or Doppler imaging of the ovaries. The only biomarker that has been extensively studied for possible use in the early detection of EOC is CA125, a high-molecular-weight glycoprotein of unknown function (Dorum et al. 1996; Fures et al. 1999). Although CA-125 has good utility for monitoring effects of treatment and for recurrence of ovarian cancer, its effectiveness as a screening tool is limited. A systematic review of the performance of the multimodal strategies of CA125 and ultrasound indicated that approximately 50% (95% confidence interval [CI] 23; 77) and 75% (95% CI 35; 97) of patients were diagnosed at stage I in CA-125-based and ultrasound screening studies, respectively (Reviews 2003). Unfortunately, the positive predictive values (PPV) of these strategies for the early detection of EOC using these modalities have been consistently less than 10% (Reviews 2003; van Nagell et al. 2000). Attempts to improve the PPV of these early detection strategies in EOC have met with limited success. These include the utilization of complex longitudinal algorithms for CA125 (McIntosh et al. 2002; Skates et al. 1995, 2003; Zhang et al. 1999), sequential testing (Berek and Bast 1995; Jacobs et al. 1999) and the addition of newer markers such as OVX-1 (Bast et al. 1983), M-CSF (Suzuki et al. 1993), lysophosphatidic acid (Xu et al. 1998), osteopontin (Kim et al. 2002) and Kallikrein 11 (McIntosh et al. 2007). In light of these considerations, novel approaches are needed for the early detection of EOC.

3 NMR-Based Metabonomics for the Analysis of Biofluids

An alternative approach for early detection of EOC is to utilize a novel and unique strategy that provides a coherent perspective of the complete metabolic response of organisms to pathophysiological insult or genetic modification (Nicholson et al. 1999). This approach to the study of metabolic processes in biological systems has been termed metabonomics (Nicholson et al. 1999) and is the focus of this paper. We have hypothesized that the analysis of a global view of metabolites in serum would enhance the possibility of identifying metabonomic signatures for EOC. Metabonomics is based on the use of NMR (and other spectroscopic methods) and multivariate statistics for biochemical data generation and interpretation. NMR spectroscopy is based on using nu-

Cancer Diagnostics Using ^{1}H-NMR-Based Metabonomics

clear spins to probe their chemical environment when placed in a static external magnetic field. Nuclei with a spin quantum number 1/2 are best suited and include ^{1}H, ^{13}C, ^{15}N and ^{31}P. Since protons are present in almost all metabolites in body fluids, a ^{1}H-NMR spectrum allows the simultaneous detection and quantification of thousands of proton-containing, low-molecular-weight species within a biological matrix, resulting in the generation of an endogenous profile that may be altered in disease to provide a characteristic fingerprint (Lindon et al. 1999, 2000; Nicholson et al. 1999, 2002).

There are several advantages of NMR-based metabonomics in a clinical setting. It can be performed on standard preparations of serum, plasma, saliva or urine, circumventing the need for specialist preparations of cellular RNA and protein required for genomics and proteomics, respectively (Lindon et al. 2000, 2001; Nicholson and Wilson 1989; Holmes et al. 2001). Moreover, since cancer is now known to be a product of the tumor–host microenvironment (Liotta and Kohn 2001), the organ-specific milieu can generate, and enzymatically modify, multiple proteins, peptides, metabolites, and cleavage products at much higher concentrations than for molecules derived only from the tumor cells.

4 ^{1}H-NMR Analysis of Plasma and Cancer Detection

The initial report indicating that ^{1}H-NMR spectroscopy of plasma might be useful for cancer detection was published in 1986 by Fossel et al. (Fossel et al. 1986). The report was based on the measurements of ^{1}H-NMR spectra of plasma samples (at either 360 or 400 MHz ^{1}H resonance frequency and at 20–22°C) for 331 subjects, including controls, patients with various types of malignant and benign tumors and pregnant women, and examination of the spectra by applying a parameter, Fossel Index (FI) which is calculated as a mean of the approximate widths at half-height of the methylene and methyl resonance envelopes. Although it appeared possible to clearly and reliably distinguish between normal controls (FI $= 39.5 \pm 1.6$ Hz) and patients with malignant tumors (FI $= 29.9 \pm 2.5$ Hz), in many subsequent studies, a remarkable overlap between cancer patients and controls was noted. This led to an intensive interlaboratory evaluation of the reproducibility and accuracy of the NMR human blood test for cancer by Chmurny et al. (Chmurny et al. 1988). This test was found to be reproducible but not accurate for screening a general asymptomatic population for cancer.

There are several limitations of these early studies. First, affected subjects in these studies had cancer of different organ sites and histologies. Clearly, there is great variability in the biology, invasiveness and metastatic potential of different tumors, and it would be surprising to find a single test that could reliably detect all or even a large number of cancers (Chmurny et al. 1988). Secondly, the predictive value of a positive screening test for cancer needs to be considered not only in the context of sensitivity and specificity, but also the prevalence of disease in the population (for example, the prevalence of EOC is relatively low). Finally, and most importantly, the early NMR studies are different from metabonomics because of the significant improvements in high-resolution NMR technology and novel computationally intense and robust analytic methods for ^{1}H-NMR spectroscopic data interpretation. These approaches, previously unavailable, have opened new avenues for disease diagnosis and management, as evidenced by the recent successful application of metabonomics to coronary heart disease (Brindle et al. 2002) and hypertension (Brindle et al. 2003), and in our studies of women with ovarian cancer.

5 ^{1}H-NMR-Based Metabonomics for Ovarian Cancer Detection

We recently conducted a study to evaluate the utility of using ^{1}H-NMR-based metabonomic analysis to discriminate samples from women with EOC from healthy controls, and women with benign ovarian diseases (Odunsi et al. 2005). Preoperative serum samples of 38 patients with EOC undergoing surgery at the Roswell Park Cancer Institute (RPCI) were collected under an approved institutional review board (IRB) protocol. The stage distribution of the EOC patients were as follows: stage I: two patients; stage IIIC: 34 patients; stage IV: two patients. Among patients with advanced disease (stages IIIC and IV), four (11%) had normal preoperative serum CA125 levels (<35 units/ml). In addition,

Cancer Diagnostics Using ^{1}H-NMR-Based Metabonomics

preoperative CA125 was normal in one of the two patients with stage I disease. The age range of the study patients was 46–86 years. For controls, the sera of 53 normal healthy women (pre- and postmenopausal controls), and 12 patients with benign ovarian cysts were collected under two additional IRB protocols at RPCI. The age range of the healthy premenopausal controls was 22–44 years, while the remaining 32 postmenopausal controls had an age range of 45–75 years. The age range of the 12 patients with benign ovarian cysts was 22–68.There were no significant differences between the study subjects and postmenopausal controls with respect to age, parity and use of oral contraceptives. Aliquots of serum were stored at –80°C until assayed.

5.1 ^{1}H-NMR Spectroscopic Analysis of the Serum Samples

Samples (100 µl) were diluted with solvent solution (99.9% D_2O) (450 µl) in 5-mm precision NMR tubes (Norell, Inc., Landisville, NJ, USA). Conventional ^{1}H-NMR spectra of the serum samples were measured at 600.22 MHz on a Bruker AMX-600 spectrometer (Billerica, MA) operating at 600 MHz ^{1}H frequency, using the pulse sequence: RD–90°–t_1–90°–t_m–90°, acquire free induction decay (FID) (i.e., the NOESYPR1D pulse sequence). RD represents a relaxation delay of 1.5 s during which the water resonance is selectively irradiated, and t_1 corresponds to a fixed interval of 4 µs. The water resonance is irradiated for a second time during the mixing time (t_m, 100 ms). For each sample, 128 FIDs were collected into 64 k data points using a spectral width of 12.2 kHz and an acquisition time of 2.69 s. The FIDs were multiplied by an exponential weighting function corresponding to a line broadening of 0.25 Hz before Fourier transformation. The acquired NMR spectra were corrected for phase and baseline distortions using UXNMR (version 97) and referenced to lactate ($CH_3\delta1.33$). Chemical components were assigned to the spectra on the basis of previously published data (Ala-Korpela 1995; Nicholson et al. 1995). Figure 1a shows the 600-MHz ^{1}H-NMR spectra of serum from a postmenopausal patient with stage 1 EOC, Fig. 1b shows the spectra from a healthy postmenopausal patient, Fig. 1c shows the spectra from a healthy premenopausal patient, while Fig. 1d shows the spectra from a patient with ovarian endometriosis. In order to remove any ambiguity in assigned chemical shift values,

Fig. 1. Comparison of patients with EOC with healthy subjects. The 600-MHz ^{1}H-NMR spectra of serum samples from a postmenopausal stage 1 EOC patient (**a**), a premenopausal healthy subject (**b**), a postmenopausal healthy subject (**c**), and a patient with benign ovarian cyst (endometriosis) (**d**). The chemical shifts of a selection of major metabolites are indicated, based on comparison with published metabolites (Pretsch et al. 1989)

samples were spiked with a small amount of three reference compounds to test whether perfect superposition of the signals could be achieved. A sample of alanine was added first, followed by valine, and then glucose with spectra acquired after each addition. In each case, the resonances of the reference fell directly on top of the assigned resonances in the biofluid.

5.2 Data Reduction of NMR Data

The ^{1}H-NMR spectra ($\delta 10$–0.2) were automatically data-reduced to 200–250 integral segments of equal length ($\delta 0.04$) using NutsPro (version 20021122, Acorn NMR, Inc., Livermore, CA, USA). Each segment consisted of the integral of the NMR region to which it was associated. To remove the effects of variation in the suppression of the water resonance, the region $\delta 5.5$ to 4.75 was set to zero integral. The data were normalized to total spectral area and centered scaling was applied.

5.3 Pattern Recognition Analysis of the ^{1}H-NMR Spectra

Principal component analysis (PCA) is an unsupervised method (i.e., analysis performed without use of knowledge of the sample class) that reduces the dimensionality of the data input while expressing much of the original n-dimensional variance in a 2D or 3D map (Eriksson et al. 1999). Prior to PCA analysis, all NMR data were mean-centered and pareto-scaled (Wold et al. 1998) to give each variable a variance numerically equal to its standard deviation. PCA was carried out on the ^{1}H-NMR data from the sera of EOC patients and controls to plot data in order to indicate relationships between samples in the multidimensional space. The principal components were displayed as a set of

Cancer Diagnostics Using ¹H-NMR-Based Metabonomics

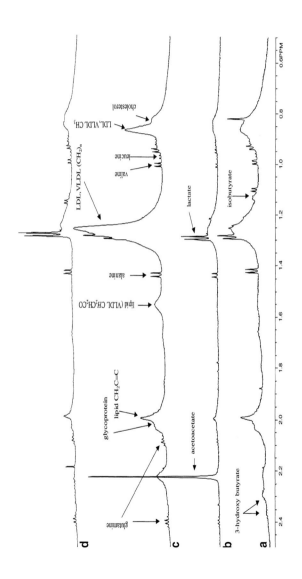

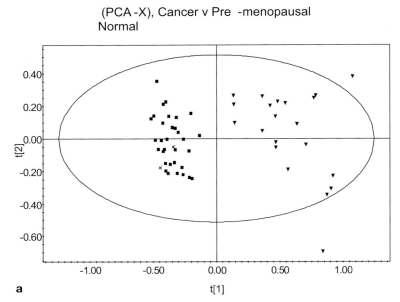

Fig. 2a. PCA plots of factor scores for the first two principal components (t[1], t[2]) showing the considerable separation achieved between (**A**) EOC serum samples (X, ■) and healthy premenopausal controls (▼)

scores (t), which highlight clustering or outliers, and a set of loadings (p), which highlight the influence of input variables on t. This data set of NMR spectra displayed good discrimination between EOC patients and controls. Thus, we were able to correctly separate all of the 38 cancer specimens (100%) and all of the 21 premenopausal normal samples (100%) (Fig. 2a). In addition, it was possible to correctly separate 37 of 38 (97.4%) cancer specimens and 31 of 32 (97%) postmenopausal control serum specimens (Fig. 2b). When patients with benign ovarian disease were included in the PCA analysis, it was still possible to correctly separate all of 38 cancer specimens (100%) from the sera of all 12 patients with benign ovarian disease (Fig. 2c). Although sera from patients with benign disease overlapped with sera from the healthy controls, it was possible to achieve separation of cancer versus noncancer

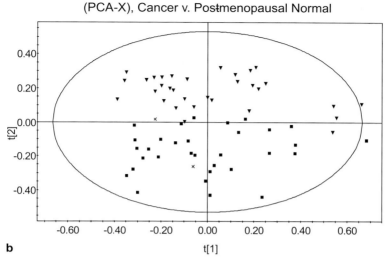

Fig. 2b. PCA plots of factor scores for the first two principal components (t[1], t[2]) showing the considerable separation achieved between EOC serum samples (X, ■) and healthy postmenopausal controls (▼)

cases. All PCA plots indicated that most of the variation occurred in the first two principal components.

5.4 Soft Independent Modeling of Class Analogy

In order to provide validation of the results, a supervised analysis of the data was performed based on soft independent modeling of class analogy (SIMCA). Since the majority of EOC patients in our study and in clinical practice are postmenopausal, we chose to perform further analysis by comparing the benign and cancer patients with healthy postmenopausal controls. SIMCA utilizes the features of PCA to construct significance limits for specified classes of samples in the scores and the residual direction. Mapping of unknown samples onto the calculated models provides the class identity based on similarity between the unknown samples and the samples in the predefined class models. A method of visualizing the SIMCA approach is the Cooman's plot

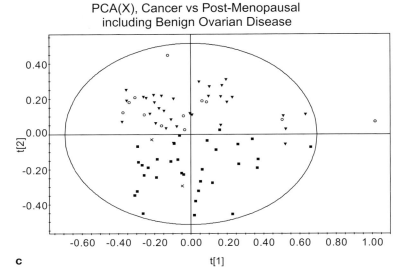

Fig. 2c. PCA plots of factor scores for the first two principal components (t[1], t[2]) showing the considerable separation achieved between EOC serum samples (X, ■), healthy postmenopausal controls (▼) and benign ovarian cysts (○). Note that optimum separation occurred in the second principle component. Patients with stage I EOC are denoted by X

(Coomans et al. 1984), which plots class distances against each other. We built separate PCA models for the sera of EOC patients and postmenopausal healthy controls. SIMCA was then applied to the models using the Cooman's plot and the classification performance was assessed by predicting class membership in terms of distance from the model. The critical distance from the model used corresponded to a 0.05 level, and defined a 95% tolerance interval. The resulting Cooman's plot demonstrated that sera classes from patients with EOC, benign ovarian cysts and the postmenopausal healthy controls did not share multivariate space, providing validation for the class separation (Fig. 3). Therefore, it should be possible to predict whether future samples can be classified as cancer or noncancer. This preliminary data demonstrated that ^{1}H-NMR-based metabonomic analysis of serum samples could achieve

Cancer Diagnostics Using ¹H-NMR-Based Metabonomics 217

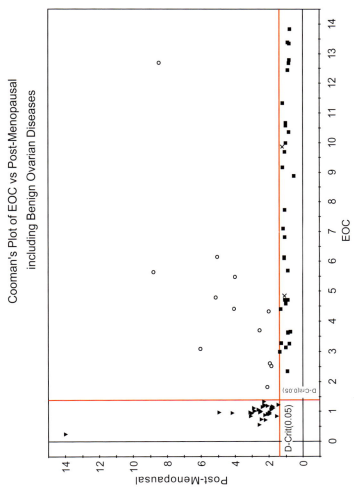

Fig. 3. Cooman's plot demonstrating that the EOC sera class (X, ■), sera from patients with benign ovarian disease (○), and the postmenopausal control sera (▼) class do not share multivariate space. Patients with stage I EOC are denoted by *X*

a clinically useful performance for the identification of serum samples of patients with EOC.

5.5 Receiver Operating Characteristic Curve Analysis

Univariate receiver operating characteristic curve (ROC) analyses were carried out via individual logistic regressions for each of 219 ^{1}H-NMR regions in order to examine their utility for predicting EOC. The sensitivity and specificity trade-offs were summarized for each variable using the area under the ROC curve denoted AUC, and calculated using the trapezoidal rule. An AUC value of 1.0 corresponds to a prediction model with 100% sensitivity and 100% specificity, while an AUC value of 0.5 corresponds to a poor predictive model (see Pepe et al. 1997 for an overview of ROC analyses via logistic regression modeling). The best two variable models were then fit starting from the univariate information via a forward stepwise selection using the AUC as the criteria for a variable's entry into the model. The data showed that a two variable model consisting of ^{1}H-NMR regions 2.77 ppm from the origin and 2.04 ppm from the origin provided a perfect fitting model, i.e., AUC = 1.0. A scatterplot is provided in Fig. 4, which clearly illustrates the delineation between the two groups. Of note, the univariate model that considered only region 2.04 ppm gave an AUC = 0.942, while the AUC for the univariate model for region 2.77 ppm gave an AUC = 0.689, i.e., prediction based upon region 2.04 is enhanced conditional upon the information contained in region 2.77 ppm. We hypothesize that the preliminary information that we have derived from this ROC analysis will allow us to refine this model for early-stage EOC, and that this approach could represent a novel strategy for the early detection of EOC.

5.6 Analysis of Spectral Pattern Differences

Based on the promising results showing complete separation of patients with EOC and controls using unsupervised PCA, supervised SIMCA, and ROC analyses applied to ^{1}H-NMR spectra of sera, we have proceeded to identify the molecules responsible for the differences in spectral patterns utilizing a previously described methodology (Gavaghan

Cancer Diagnostics Using ¹H-NMR-Based Metabonomics

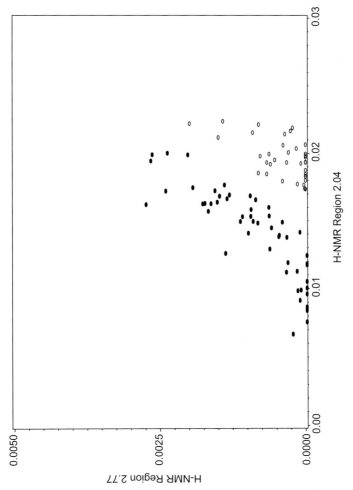

Fig. 4. Scatter plot and ROC analysis of H-NMR metabonomic profile of sera from healthy postmenopausal controls (*solid circles*) and EOC patients (*empty circles*)

et al. 2000). The regions of the NMR spectrum that most strongly influence separation between EOC and healthy controls are indicated by the regression coefficients (data not shown). The coefficients were derived from the PCA models and each bar represents a spectral region covering 0.04 ppm, showing how the ^{1}H-NMR profile of the EOC samples differed from the ^{1}H-NMR profile of the healthy serum samples. A negative value indicates a relatively greater concentration of metabolite (assigned using NMR chemical shift assignment tables) present in EOC samples and a positive value indicates a relatively lower concentration, with respect to EOC samples. In general, the regression coefficients, or loadings, most influential for the EOC samples compared with postmenopausal controls lie around δ3.7 ppm (due to various sugar hydrogens) while the loadings most influential for the EOC samples compared with premenopausal controls lie around δ2.25 (due to acetoacetate). Other loadings suggest greater amounts of 3-hydroxybutyrate and isobutyrate in the sera of EOC patients compared with pre- and postmenopausal controls. These differences are also readily apparent on visual inspection of the spectra. The biological significance of these observations is currently unclear.

5.7 Validation of EOC Results in an Independent Set of Serum Specimens

In an effort to validate the results described above, we recently examined an independent set of specimens from the Databank and Biorepository core facility at Roswell Park Cancer Institute. We compared 20 patients with stage III ovarian cancer with age-matched controls. The results confirm our original data and indicate considerable separation of EOC patients from healthy controls by PCA and PLS-DA methods (Fig. 5a and b).

5.8 Mass Spectrometry-Based Metabolic Profiling in Ovarian Cancer

In a recent study, the combination of gas chromatography/time-of-flight mass spectrometry (GC-TOF MS) was used to analyze 66 invasive ovarian carcinomas and nine borderline tumors of the ovary (Denkert et al. 2006). After automated mass spectral deconvolution, 291 metabolites were detected, of which 114 (39.1%) were annotated as known com-

Cancer Diagnostics Using ¹H-NMR-Based Metabonomics

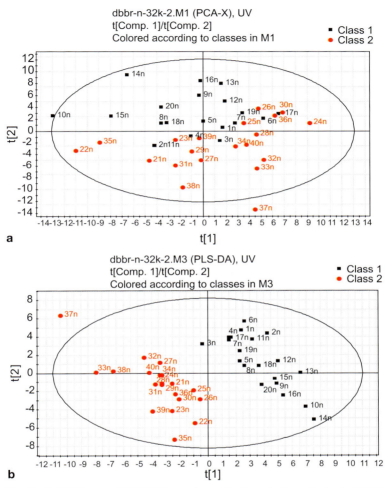

Fig. 5a, b. Validation of 1H-NMR metabonomics in an independent set of EOC specimens. **a** PCA plot and **b** PLS-DA plot showing considerable separation between EOC cases (*red*) vs healthy controls (*black*)

pounds. Principal component analysis (PCA) revealed four principal components that were significantly different between the two groups, with the highest significance found for the second component ($p = 0.00000009$). PCA as well as additional supervised predictive models allowed a separation of 88% of the borderline tumors from the carcinomas. Using the KEGG database, the authors linked metabolic changes to putative key enzymes that play an important role in the corresponding pathways. These include enzymes that regulate pyrimidine metabolism such as dihydropyrimidine dehydrogenase and thymidine phosphorylase. The unique aspect of the study is the demonstration that metabolic profiling using GC-TOF MS is suitable for analysis of fresh frozen human tumor samples. Moreover, there appears to be a consistent and significant change in primary metabolism of ovarian tumors, which can be detected using multivariate statistical approaches.

In another recent study, peptides were extracted from frozen tissues of 25 ovarian carcinomas (stages III and IV) and 23 benign ovaries; and analyzed using MALDI-TOF MS, nanoESI MS and MS/MS (Lemaire et al. 2007). A marker with anm/z of 9744 corresponding to 84 amino acid residues from the 11S proteasome activator complex (PA28 or Reg-alpha) was identified. This marker was subsequently validated using MALDI imaging, classical immunocytochemistry with an antibody raised against the C-terminal part of the protein, and Western blot analysis. Together these two studies (Denkert et al. 2006; Lemaire et al. 2007) indicate that direct tissue analysis by mass spectrometry-based strategies can facilitate biomarker discovery and validation in human cancers.

6   Conclusions and Future Directions

There have been remarkable efforts by several groups of investigators to identify reliable markers for early detection of a wide range of solid tumors, including epithelial ovarian cancer (EOC). High-throughput metabolite profiling and protein expression analysis aimed at the identification of metabolites that are generated as a consequence of tumor–host interaction could provide a strategy for (a) discriminating cancer cases from healthy control candidates and (b) identifying a panel of metabolites that could be useful as biomarkers of early detection and targets of therapy. The rapid development of metabonomics and proteomics-based technologies are bringing new perspectives that will likely integrate both approaches leading to a comprehensive and global view of cancer cell behavior. Widespread and routine use of metabonomics for cancer diagnosis will require the implementation of carefully developed SOPs based on larger studies in various cancer types. The identification of a repertoire of metabolites and proteins that mark the transition from normal to the transformed phenotype should allow detection of cancer at a preclinical stage, where the chances for cure would be highest.

Acknowledgements. I thank Dr. Thomas Szyperski for helpful discussions. Supported by the Oshei Foundation, Roswell Park Cancer Center Support Grant P30CA16056 and R21 CA106949–01A1 from the National Institutes of Health.

References

Ala-Korpela M (1995) ^{1}H-NMR spectroscopy of human blood plasma. Prog Nucl Magn Reson Spectrosc 27:475–554

Armstrong DK, Bundy B, Wenzel L, Huang HQ, Baergen R, Lele S, Copeland LJ, Walker JL, Burger RA (2006) Intraperitoneal cisplatin and paclitaxel in ovarian cancer. N Engl J Med 354:34–43

Bast RC Jr, Klug TL, St John E, Jenison E, Niloff JM, Lazarus H, Berkowitz RS, Leavitt T, Griffiths CT, Parker L, Zurawski VR Jr, Knapp RC (1983) A radioimmunoassay using a monoclonal antibody to monitor the course of epithelial ovarian cancer. N Engl J Med 309:883–887

Berek JS, Bast RC Jr (1995) Ovarian cancer screening. The use of serial complementary tumor markers to improve sensitivity and specificity for early detection. Cancer 76:2092–2096

Brindle JT, Antti H, Holmes E, Tranter G, Nicholson JK, Bethell HW, Clarke S, Schofield PM, McKilligin E, Mosedale DE, Grainger DJ (2002) Rapid and noninvasive diagnosis of the presence and severity of coronary heart disease using 1H-NMR-based metabonomics. Nat Med 8:1439–1444

Brindle JT, Nicholson JK, Schofield PM, Grainger DJ, Holmes E (2003) Application of chemometrics to 1H NMR spectroscopic data to investigate a relationship between human serum metabolic profiles and hypertension. Analyst 128:32–36

Chmurny GN, Hilton BD, Halverson D, McGregor GN, Klose J, Issaq HJ, Muschik GM, Urba WJ, Mellini ML, Costello R et al (1988) An NMR blood test for cancer: a critical assessment. NMR Biomed 1:136–150

Coomans D, Broeckaert I, Derde MP, Tassin A, Massart DL, Wold S (1984) Use of a microcomputer for the definition of multivariate confidence regions in medical diagnosis based on clinical laboratory profiles. Comput Biomed Res 17:1–14

Denkert C, Budczies J, Kind T, Weichert W, Tablack P, Sehouli J, Niesporek S, Konsgen D, Dietel M, Fiehn O (2006) Mass spectrometry-based metabolic profiling reveals different metabolite patterns in invasive ovarian carcinomas and ovarian borderline tumors. Cancer Res 66:10795–10804

Dorum A, Kristensen GB, Abeler VM, Trope CG, Moller P (1996) Early detection of familial ovarian cancer. Eur J Cancer 32A:1645–1651

Eriksson L, Johansson E, Kettaneh-Wold N, Wold S (1999) Introduction to multi- and megavariate data analysis using projection methods (PCA, PLS) Umetrics, Umea, Sweden

Fossel ET, Carr JM, McDonagh J (1986) Detection of malignant tumors. Water-suppressed proton nuclear magnetic resonance spectroscopy of plasma. N Engl J Med 315:1369–1376

Fures R, Bukovic D, Hodek B, Klaric B, Herman R, Grubisic G (1999) Preoperative tumor marker CA125 levels in relation to epithelial ovarian cancer stage. Coll Antropol 23:189–194

Gavaghan CL, Holmes E, Lenz E, Wilson ID, Nicholson JK (2000) An NMR-based metabonomic approach to investigate the biochemical consequences of genetic strain differences: application to the C57BL10J, Alpk:ApfCD mouse. FEBS Lett 484:169–174

Greenlee RT, Hill-Harmon MB, Murray T, Thun M (2001) Cancer statistics, 2001. CA Cancer J Clin 51:15–36

Holmes E, Nicholson JK, Tranter G (2001) Metabonomic characterization of genetic variations in toxicological and metabolic responses using probabilistic neural networks. Chem Res Toxicol 14:182–191

Jacobs IJ, Skates SJ, MacDonald N, Menon U, Rosenthal AN, Davies AP, Woolas R, Jeyarajah AR, Sibley K, Lowe DG, Oram DH (1999) Screening for ovarian cancer: a pilot randomised controlled trial. Lancet 353:1207–1210

Kim JH, Skates SJ, Uede T, Wong KK, Schorge JO, Feltmate CM, Berkowitz RS, Cramer DW, Mok SC (2002) Osteopontin as a potential diagnostic biomarker for ovarian cancer. JAMA 287:1671–1679

Lemaire R, Menguellet SA, Stauber J, Marchaudon V, Lucot JP, Collinet P, Farine MO, Vinatier D, Day R, Ducoroy P, Salzet M, Fournier I (2007) Specific MALDI imaging and profiling for biomarker hunting and validation: fragment of the 11S proteasome activator complex reg alpha fragment is a new potential ovary cancer biomarker. J Proteome Res 6:4127–4134

Cancer Diagnostics Using ^{1}H-NMR-Based Metabonomics 225

Lindon JC, Nicholson JK, Everett JR (1999) NMR spectroscopy of biofluids. Annu Rep NMR Spectrosc 38:1–88

Lindon JC, Nicholson JK, Holmes E, Everett JR (2000) Metabonomics: metabolic processes studied by NMR spectroscopy of biofluids. Concepts Magn Reson 12:289–320

Lindon JC, Holmes E, Nicholson JK (2001) Pattern recognition methods and applications in biomedical magnetic resonance. Prog Nucl Magn Reson Spectrosc 39:1–40

Liotta LA, Kohn EC (2001) The microenvironment of the tumour-host interface. Nature 411:375–379

McGuire WP, Hoskins WJ, Brady MF, Kucera PR, Partridge EE, Look KY, Clarke-Pearson DL, Davidson M (1996) Cyclophosphamide and cisplatin versus paclitaxel and cisplatin: a phase III randomized trial in patients with suboptimal stage III/IV ovarian cancer (from the Gynecologic Oncology Group). Semin Oncol 23:40–47

McIntosh MW, Urban N, Karlan B (2002) Generating longitudinal screening algorithms using novel biomarkers for disease. Cancer Epidemiol Biomarkers Prev 11:159–166

McIntosh MW, Liu Y, Drescher C, Urban N, Diamandis EP (2007) Validation and characterization of human kallikrein 11 as a serum marker for diagnosis of ovarian carcinoma. Clin Cancer Res 13:4422–4428

Menon U, Jacobs IJ (2000) Recent developments in ovarian cancer screening. Curr Opin Obstet Gynecol 12:39–42

Nicholson JK, Wilson ID (1989) High resolution proton magnetic resonance spectroscopy of biological fluids. Prog Nucl Magn Reson Spectrosc 21:449–501

Nicholson JK, Foxall PJ, Spraul M, Farrant RD, Lindon JC (1995) 750 MHz 1H and 1H-13C NMR spectroscopy of human blood plasma. Anal Chem 67:793–811

Nicholson JK, Lindon JC, Holmes E (1999) 'Metabonomics': understanding the metabolic responses of living systems to pathophysiological stimuli via multivariate statistical analysis of biological NMR spectroscopic data. Xenobiotica 29:1181–1189

Nicholson JK, Connelly J, Lindon JC, Holmes E (2002) Metabonomics: a platform for studying drug toxicity and gene function. Nat Rev Drug Discov 1:153–161

Odunsi K, Wollman RM, Ambrosone CB, Hutson A, McCann SE, Tammela J, Geisler JP, Miller G, Sellers T, Cliby W, Qian F, Keitz B, Intengan M, Lele S, Alderfer JL (2005) Detection of epithelial ovarian cancer using 1H-NMR-based metabonomics. Int J Cancer 113:782–788

Pepe MS (1997) A regression modeling framework for receiver operating characteristic curves in medical diagnostic testing. Biometrika 84:595–608

Pretsch E, Seibl J, Simon W, Clerc T (1989) Spectral data for structure determination of organic compounds. Springer-Verlag, Berlin Heidelberg New York

Reviews E (2003) Database of abstracts of reviews of effects NHS Centre for Reviews and Dissemination. Screening for ovarian cancer. Database of Absracts of Reviews of Effectiveness

Skates SJ, Xu FJ, Yu YH, Sjovall K, Einhorn N, Chang Y, Bast RC Jr, Knapp RC (1995) Toward an optimal algorithm for ovarian cancer screening with longitudinal tumor markers. Cancer 76:2004–2010

Skates SJ, Menon U, MacDonald N, Rosenthal AN, Oram DH, Knapp RC, Jacobs IJ (2003) Calculation of the risk of ovarian cancer from serial CA-125 Values for preclinical detection in postmenopausal women. J Clin Oncol 21:206–210

Suzuki M, Ohwada M, Aida I, Tamada T, Hanamura T, Nagatomo M (1993) Macrophage colony-stimulating factor as a tumor marker for epithelial ovarian cancer. Obstet Gynecol 82:946–950

van Nagell JR Jr, DePriest PD, Reedy MB, Gallion HH, Ueland FR, Pavlik EJ, Kryscio RJ (2000) The efficacy of transvaginal sonographic screening in asymptomatic women at risk for ovarian cancer. Gynecol Oncol 77:350–356

Wingo PA, Tong T, Bolden S (1995) Cancer statistics, 1995. CA Cancer J Clin 45:8–30

Wold S, Antti H, Lindgren F, Ohman J (1998) Orthogonal signal correction of near-infrared spectra. Chemom Intell Lab Syst 44:175–185

Xu Y, Shen Z, Wiper DW, Wu M, Morton RE, Elson P, Kennedy AW, Belinson J, Markman M, Casey G (1998) Lysophosphatidic acid as a potential biomarker for ovarian and other gynecologic cancers. JAMA 280:719–723

Young RC, Walton LA, Ellenberg SS, Homesley HD, Wilbanks GD, Decker DG, Miller A, Park R, Major F Jr (1990) Adjuvant therapy in stage I and stage II epithelial ovarian cancer. Results of two prospective randomized trials. N Engl J Med 322:1021–1027

Zhang Z, Barnhill SD, Zhang H, Xu F, Yu Y, Jacobs I, Woolas RP, Berchuck A, Madyastha KR, Bast RC Jr (1999) Combination of multiple serum markers using an artificial neural network to improve specificity in discriminating malignant from benign pelvic masses. Gynecol Oncol 73:56–61

Ernst Schering Foundation Symposium Proceedings, Vol. 4, pp. 227–249
DOI 10.1007/2789_2008_096
© Springer-Verlag Berlin Heidelberg
Published Online: 03 July 2008

Human Metabolic Phenotyping and Metabolome Wide Association Studies

E. Holmes[✉], J.K. Nicholson

Divsion of Surgery, Oncology, Reproductive Biology and Anaesthetics (SORA),
Faculty of Medicine, Sir Alexander Fleming Building, South Kensington,
SW7 2AZ London, UK
email: *Elaine.holmes@imperial.ac.uk*

1	Introduction	228
2	Defining the "Normal" Phenotype	229
3	Detecting Pathophenotypes: Diagnostics	232
3.1	Metabolic Profiling of Insulin Resistance	232
3.2	Cardiovascular Disease	234
3.3	Metabolic Investigations of Neuropathological Disease	235
3.4	Intestinal Disorders	236
3.5	Cancers	237
3.6	Infectious Diseases	237
4	Defining Biomarkers	238
5	The Way Forward	239
References		242

Abstract. Metabolic phenotyping in large-scale population studies can yield crucial information regarding the impact and interaction of genetic and environmental factors with regard to the prevalence and risk of chronic diseases. Spectroscopic technologies such as nuclear magnetic resonance (NMR) spectroscopy and mass spectrometry (MS) can be used to generate multi-parameter profiles of biological samples and together with automated sample delivery and mathematical modelling systems, can be used as a high throughput screening tool. The adaptation of these metabolic profiling tools from pre-clinical studies in animal models to population studies in man is explored and an overview of the current and future roles of metabolic phenotyping is described, including the idea of "Metabolome Wide Association Screening" focussing on key disease areas such as cardiovascular disease and metabolic syndrome, cancers and neurodegeneration.

1 Introduction

Recognition of the inadequacy of the genome sequence to explain the fundamental nature of many disease processes has precipitated a marked increase in the evaluation of approaches that relate gene expression to phenotypic outcomes. There is also increasing recognition of biological complexity and the conceptual paradigm has been shifted from simple univariate measurements of response to the need to integrate technologies and their outputs in order to operate at a systems biology level. Interactions of genes, proteins and metabolites at different levels of biomolecular organization can be probed by various technologies and integrated using bioinformatic and chemometric strategies to extract latent information that carries a diagnostic or even prognostic signature. One of the major goals of twenty-first century medicine will be the introduction of personalized health care through a holistic understanding of an individual's overall biochemical status. In order to achieve this aim, the effects of both genetic predisposition and a wide range of environmental factors such as diet, drug intake, smoking habits, stress and amount of physical activity, etc., need to be taken into account. Metabonomics (variously referred to as metabolomics or metabolic profiling) (Nicholson et al. 1999, 2002; Fiehn et al. 2000) is a rapidly emerging field of research combining sophisticated analytical tools such as nuclear magnetic resonance (NMR) spectroscopy and mass spectrometry with multivariate statistical analysis to generate complex metabolic profiles of biofluids and tissues. Pathological stimuli or genetic modification influence metabolite profiles in a characteristic and consistent manner, involving adjustment of the intra- and extracellular fluids as the organism strives to maintain homeostatic equilibrium. By harnessing appropriate mathematical and pattern recognition procedures to interrogate the data produced by high-resolution spectral analysis, characteristic profiles of physiological or pathological responses can be established.

Human Metabotype Profiling 229

The global objective of this chapter is to review the potential of metabolic profiling methods for characterizing the complex metabolic phenotype of humans in health and disease. Metabolic profiling has been successfully applied across a wide range of fields in plant and animal biology such as characterization of natural products (Bailey et al. 2002), monitoring response to therapeutic or nutritional interventions (Neild et al. 1997; Lamers et al. 2003; Wang et al. 2004), toxicology (Ebbels et al. 2007), drug metabolism (Foxall et al. 1996; Plumb et al. 2003), functional genomics (Gavaghan et al. 2000) and disease diagnosis and prognosis (Brindle et al. 2002; Yi et al. 2006; Clayton et al. 2006). The vast majority of metabolic profiling studies have been conducted in laboratory models of disease or toxicity where control over genetic and environmental conditions can be exercised. However, given the substantial array of animal studies that identify the metabolic response to controlled interventions, it is now appropriate to expand the available knowledge to address more complex phenotypes and, in particular, to extend the methodology to investigate human metabolism. The potential of metabolic profiling to address complex human clinical and even epidemiological questions has vastly increased due to recent advances in both analytical and mathematical technology; including capacity for higher throughput of samples, increased analytical sensitivity and the evolution of mathematical methods for accommodating analytical and biological variation. In this chapter, illustrations of research where metabolic profiling has already been employed in investigating human health and disease is summarized, and potential areas which would benefit from application of such technology are outlined.

2 Defining the "Normal" Phenotype

Prior to utilizing metabolic profiling technology for diagnostic purposes in human studies, it is first necessary to define the metabolic range covered by normal physiological variation. Only then can robust and specific biomarkers of disease be extracted. Metabolic variation is dependent on both genetic and environmental parameters, and each biological tissue or fluid has its own unique metabolic signature (Fig. 1). Ethnicity, gender, age, activity, nutritional status, medication, stress, polymor-

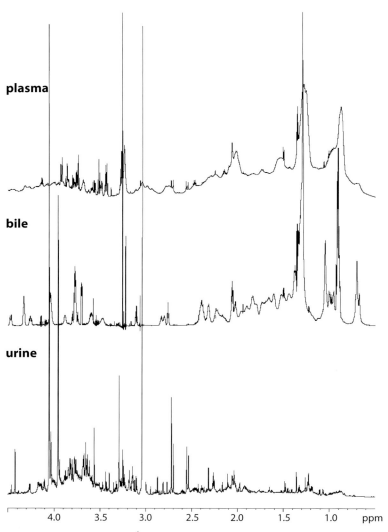

Fig. 1. Standard 600 MHz ^{1}H NMR spectra showing characteristic profiles for urine, plasma and bile

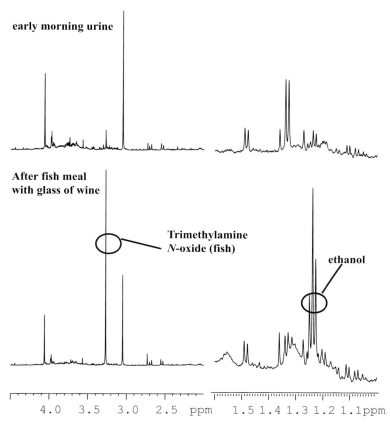

Fig. 2. Selected regions of standard 600-MHz ^{1}H NMR spectrum of urine samples from a healthy male before and after consuming an evening meal showing characteristic metabolic changes associated with the consumption of fish and a glass of wine (vertical scale for region on right hand side of plot x5)

phisms, hormone levels and circadian cycles are all known to impact upon mammalian metabolite profiles (Holmes et al. 1994; Slupsky et al. 2007; Williams et al. 2006; Bollard et al. 2001; Teague et al. 2004, 2006) (Fig. 2). Evaluation of normal ranges of mammalian metabolite composition under various physiological and analytical conditions can be found in the literature for several biological matrices, including urine (Holmes et al. 1994; Maher et al. 2007), plasma (Teahan et al. 2006; Lenz et al. 2003), cerebrospinal fluid (CSF) (Koschorek et al. 1993), feces (Saric et al. 2008) and various tissues (Tsang et al. 2005; Wang et al. 2008; Garrod et al. 1999). The extent of variation and the dynamic ranges of metabolite concentrations in metabolite profiles are dependent upon the influence of homeostatic mechanisms on that biological matrix (Fig. 1). For example, plasma composition is maintained under homeostatic control and metabolite concentrations are found to be relatively stable in terms of both qualitative and quantitative differences in comparison with excretory biofluids such as urine, where metabolite concentrations vary greatly in terms of both the presence and quantity.

3 Detecting Pathophenotypes: Diagnostics

In many instances, diagnosis of the presence of a disease is achievable by routine and inexpensive clinical assays or genetic tests, for example type 2 diabetes, inborn errors of metabolism such as phenylketonuria, and many neurodegenerative disorders (Guthrie and Susi 1963; International Huntington Association and the World Federation of Neurology Research Group on Huntington's Chorea 1994). However, for some diseases, early diagnosis remains the key issue, and even for those diseases that are easily diagnosed by simple assays, in some cases the stage of disease is harder to determine accurately. Therefore, improved diagnostics are required in order to establish the optimal therapeutic management. Here the application of metabolic profiling can be an efficient tool for differential diagnosis of various disease conditions, as has been shown for a wide range of diseases, including cardiovascular, intestinal disorders, cancers, renal disease (Fig. 3), osteopathies and neuropathologies. Several examples are discussed in the following sections.

3.1 Metabolic Profiling of Insulin Resistance

Insulin resistance (IR) is one of the fastest growing human pathological conditions, and is now an increasing health burden in the develop-

ANALYSIS OF RENAL FANCONI URINE BY MS AND NMR

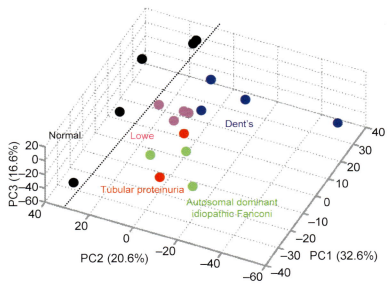

Fig. 3. Principal Components scores plot derived from the NMR and MS profiles of urine obtained from humans with different types of Fanconi syndrome. (Adapted from Vilasi 2007)

ing world as well as westernized societies. IR has been studied across a wide range of animal models using metabolic profiling and large-scale epidemiological studies are now being undertaken in human populations. Several studies on models of insulin resistance and type 2 diabetes have been undertaken in animal models. For example, the effects of streptozotocin-induced diabetes have been profiled using NMR with principal components analysis (PCA) (Nemoto et al. 2007). The effects of a high-fat diet were explored in inbred mouse strains selected for their resistance (BALB/c) and susceptibility (129S6) to IR and nonalcoholic fatty liver disease (NAFLD). High plasma concentrations of phosphocholine and increased urinary excretion of methylamines, associated with changes in gut microflora were found (Dumas et al. 2006a). Several studies have also been conducted on the Zucker rat, which is a common animal model for IR and obesity (Dumas et al. 2006b; Yi et al. 2006). Several recent EU-funded initiatives such as MolPAGE, FGENT-CARD, PROCARDIS involve or even focus on metabolic profiling of the human IR phenotype. Most of the early publications arising from these studies encompass an exploration of variation in human biofluids (Teague et al. 2004; Maher et al. 2007; Plumb et al. 2005), but several smaller studies targeting IR have identified specific metabolic phenotypes or metabotypes associated with IR and type 2 diabetes (Williams et al. 2005; Atherton et al. 2006). Indeed, type 2 diabetes was first profiled using NMR spectroscopy in 1984 (Bales et al. 1984). Although as yet there are few substantive papers exploring the more subtle and substantive metabolic consequences of IR, an explosion in the literature reporting on some of the major epidemiological studies is imminent.

3.2 Cardiovascular Disease

Like IR, cardiovascular disease (CVD) is also a part of the metabolic syndrome spectrum and is also growing at an alarming rate. CVD has been studied across several small populations using metabolic profiling approaches. Predominantly NMR spectroscopy-based studies on plasma or serum, in particular, have yielded metabolic profiles that are differentiated from control or healthy profiles in both the lipoprotein profiles, choline metabolites and in some of the lower molecular weight metabolic components (Brindle et al. 2002; Kirschenlohr et al. 2006). One LC-MS study, conducted on patients with myocardial ischaemia, some of whom demonstrated inducible ischemia and some of whom did not, was able to separate the two groups clearly on the basis of citric and lactic acid amongst other metabolites. However, of the 23 metabolites identified as candidate biomarkers, few were identified and the study lacked a matched control group. Nevertheless, the potential of LC-MS methodology to characterize myocardial ischemia was clearly demonstrated (Sabatine et al. 2005). Although early studies have produced promising results in terms of obtaining a diagnostic signature, due to confounders such as medication, the higher prevalence of the disease in men and the high dependency on diet and lifestyle, there is

Human Metabotype Profiling

still a requirement for larger-scale definitive studies in this area. Identification of an early diagnostic for CVD, or even a prognostic signature, would undoubtedly be one of the "Holy Grails" of metabolic profiling. Several population-based epidemiology studies have been designed to test hypotheses regarding the relationship between the development of hypertension, a condition that predisposes to CVD, and various lifestyle factors. In particular, the INTERMAP study (International Study of Macronutrients and Blood Pressure) was launched in 1996 to investigate the relationship of dietary intake of macronutrients and other factors to blood pressure across four countries: China, Japan, the United Kingdom and the United States (Stamler et al. 2003). This study involved collection of two 24-h urine samples from 4,680 participants, in addition to blood pressure measurements and NMR spectroscopy; the first results are beginning to emerge (Dumas et al. 2006b; Homes et al. 2007). Furthermore we have recently introduced the concept of the "Metabolome-wide association study" (Holmes et al. 2008) demonstrating broad metabolite profile screening is linked statistically to disease risk factor data to identify new molecular targets in metabolism that can be physiologically tested. Thus, whilst large epidemiological studies present a practical and logistical challenge, they are at least feasible and metabolic profiling is well suited to characterizing the metabolic phenotypes of populations, which have high risk and prevalence of pathological or prepathological conditions such as CVD and hypertension.

3.3   Metabolic Investigations of Neuropathological Disease

Disease progression in many neurodegenerative and psychological disorders is difficult to assess with batteries of cognitive or psychological tests forming part of the diagnostic for disease stage. For these pathologies, it would be ideal to have a metabolic indicator of disease stage in order to achieve the optimal therapeutic intervention strategy. In the neurodegeneration field, there are many more studies using magnetic resonance spectroscopy (MRS) of tissues than high-resolution NMR spectroscopy; however, MRS profiles lack sensitivity in comparison. Again, in experimental models such as the transgenic R6/2 mouse model of juvenile Huntington's disease (Bates et al. 1997), NMR-based metabonomic studies in the R6/2 mouse characterized the metabolic signature of HD in several tissues and body fluids (urine, plasma, skeletal muscle, striatum, cerebral cortex, cerebellum and brain stem) at 4, 8 and 12 weeks of age (Tsang et al. 2006b). This study supported previous results obtained by Jenkins et al. (1993) using MRS, but additionally was able to resolve choline and glycerophosphocholine resonances. Choline levels were observed to decrease in most of the neuroanatomical regions analysed in the R6/2 mouse, whereas glycerophosphocholine increased suggestive of a pro-catabolic phenotype in the R6/2 mouse model. This has also been shown in a small number of HD patients where glycerophosphocholine levels correlated with disease progression (Underwood et al. 2006). Currently, the European Huntington's Disease Network is focused on collaborating across European cohorts to establish biomarkers of HD and is actively employing a systems biology approach combining transcriptomic, proteomic and metabonomic data from HD patients and age-matched controls. For schizophrenia, the metabolic profiling strategy has been taken one step further, and not only has the metabolic phenotype of the pathology been defined (Tsang et al. 2006a), but a preliminary study evaluating response to therapeutic intervention with antipsychotics has been profiled, showing that those patients treated on the first episode of the disease were able to achieve normalization of their spectral profiles (Holmes et al. 2006).

3.4 Intestinal Disorders

Although it is relatively easy to diagnose irritable bowel disorders, discriminating between them, for example Crohn's disease (CD) and ulcerative colitis, can provide more of a challenge. Moreover, monitoring the condition generally involves an invasive series of surgical procedures such as colonoscopy. Recently spectroscopic methods have been applied to stool samples from CD, ulcerative colitis, polyposis and colon cancer to achieve discriminatory profiles for each of these conditions (Marchesi et al. 2007; Scanlan et al. 2008). Faecal water profiles from each of these conditions were found to have higher levels of amino acids, lower levels of short chain fatty acids and characteristic bile acid signatures, although the specific amino acids and short chain fatty acids that changed were different for each condition.

Human Metabotype Profiling 237

3.5 Cancers

Cancer is one area where a specific and distinctive metabolic profile has remained elusive. The first studies performed on profiling cancer using NMR spectroscopy as a diagnostic tool were unfortunately badly confounded (Fossel et al. 1986), which resulted in avoidance of this area for many years. Two of the main problems with cancer diagnostics are the lack of specificity and the fact that many of the metabolic changes are associated with inflammation. Now, however, with the recent advances in technology, several studies on small cohorts of patients have produced promising results. In one such study Odunsi et al. were able to differentiate between patients with ovarian cancer and matched controls using NMR analysis of blood plasma (Odunsi et al. 2005), whilst in another study, excised tumour tissue was analysed using GC-MS and ovarian cancers were differentiated from borderline tumours with high sensitivity (Denkert et al. 2006). Because of the difficulty of finding cancer-specific biomarkers, several studies have employed more than one "omics" platform. For example, renal cell carcinoma has been characterized using a combined proteomic and MS-based metabolic profiling approach (Perroud et al. 2007). Due to the obvious effect on glycolysis in tumours, studies on cancer cell lines often employ ^{13}C-labelled glucose. Using this labelling approach, characterization was achieved for a breast cancer mammary epithelial line from a normal mammary epithelial line (Yang et al. 2007) using a combination of NMR and GC-MS. Whilst such studies can potentially throw light on mechanisms and aid drug target discovery, the metabolic situation is very different and inherently more complex inside the human, and one must bear in mind the biomarkers discovered via metabolic profiling of cell lines may not always be translatable.

3.6 Infectious Diseases

There have been a number of metabolic profiling initiatives in the infectious disease area, including parasitic infection, tuberculosis and meningitis (Singer et al. 2006; Glickman et al. 1994; Coen et al. 2005). In reality, infectious diseases are predominant in developing countries and therefore relatively little metabolic profiling work has been done in this area due to financial constraints and practicality. Further complications of applying the technology in this area is the fact that multiple infection is the norm for many of these populations (Buck et al. 1978), thereby rendering extraction of a panel of biomarkers for single infection difficult, although arguably it would be preferable to profile the multiple diseases simultaneously. Since metabolic profiling is an inexpensive technology, particularly when used in an exploratory capacity with subsequent development of biomarker assays, it has great potential in the diagnosis and surveillance of infectious diseases.

4 Defining Biomarkers

In order to be truly useful, a biomarker must be quantifiable, reproducible and analytically simple to measure (Atkinson et al.). Other desirable qualities of biomarkers are that the biomarker is inexpensive to measure, its concentration or level does not vary across a large range, it is specific to the condition of interest and that it is not affected by co-morbid factors.

The capacity for metabolic profiling approaches to generate diagnostic molecular signatures has been demonstrated for a range of conditions in human studies, but many studies have been preliminary in nature and now require extensive validation across larger cohorts on individuals.

Biomarker detection plays a key role in the discovery and development of new treatments for human disease and therefore there has been a great deal of method development in the area of improving biomarker detection and extraction from large multivariate data sets.

Increased sensitivity of analytical detection is useless without the means to interpret the greater number of candidate molecules or signals generated by an analytical platform. The three major analytical platforms—GC-MS, LC-MS and NMR spectroscopy—have strengths and weaknesses that are partially determined by the nature of the disease or intervention under investigation. Although GC-MS typically requires time-consuming derivatization steps, there are several good databases for molecular identification once the data are acquired. To improve molecular identification, GC-MS data can be deconvolved using hierarchical multivariate curve resolution to resolve the spectra into pure profiles of compounds (Jonsson et al. 2006). GC-MS is well suited to measuring diseases where targeted analysis can be applied to a set of molecules which are known to carry a signature for a particular disease, for example the measurement of organic acids for characterization of many inborn errors of metabolism. LC-MS, and more particularly UPLC-MS, with its enhanced resolution and sensitivity, provides a comprehensive signature of metabolic perturbation and is becoming increasingly useful as the databases associated with molecular identification from retention-time–m/z pairs improve. NMR spectroscopy is the most reproducible of the three techniques and the least prone to artefact. Sensitivity remains lower than MS methods even with the use of cryoprobes, but the technique is inherently more amenable to structural elucidation. For a few well-funded laboratories, the obvious choice is to employ all three techniques. For the rest then, a sensible choice has to be made based on cost, laboratory infrastructure and the clinical, therapeutic or nutritional areas of interest.

Whatever the platform of choice, there is a continuous stream of new chemometric and bioinformatics processing and preprocessing techniques for optimization of the analysis of spectral data including algorithms for curve resolution (Jonsson et al. 2006), peak alignment (Jonsson et al. 2005; Csenki et al. 2007; Stoyanova et al. 2004), normalization (Dieterle et al. 2006) and quantification (Vehtari et al. 2007) in order to provide the best chance of capturing potential biomarkers. Other methods focus more on the identification of correlation within the data structure in order to provide as much information as possible regarding the identity of biomarkers, for example statistical correlation spectroscopy (Cloarec et al. 2005; Crockford et al. 2006).

5   The Way Forward

Arguably, the most valuable type of biomarker is either an early diagnostic or even prognostic, i.e. one which allows detection of a disease prior to the manifestation of clinical symptoms. Identification of prognostic biomarkers can result in prevention of the development of that pathology, or even the reversal of the pathology. Several recent studies have indicated that for certain conditions metabolic profiling can uncover a prognostic signature. For example, from the predose ^{1}H NMR urine profile, it is possible to predict animals who will develop toxicity after an oral dose of galactosamine and to predict toxicity associated with paracetamol ingestion (Clayton et al. 2006).

Psychological and physiological stress have also been shown to predispose individuals to a number of illnesses and conditions and the concept of allostatic load, the indication of wear-and-tear on multiple biological systems as they adapt and respond, within the individual, to life's demands (McEwen 2002). It was found that for men at highest risk of mortality, a cluster of five biomarkers are usually present at elevated levels, namely CRP, IL-6, fibrinogen, norepinephrine, and epinephrine (Gruenewald et al. 2006). Application of metabolic profiling strategies to epidemiological cohorts should finally give enough power to make associations between gene-gene and gene-environment factors and the associated consequence on the metabolic phenotype.

As the world turns towards systems biology, there is a pressing need to begin to integrate multiple "-omics" data sensibly. Simply making lists of genes, proteins and metabolites that are altered by a particular disease, mathematical modelling solutions can help to extract latent information and can use aspects of one "-omics" data set to strengthen another. This integration has been attempted at a preliminary level in animal models with relatively small group sizes, but has yet to be applied on a large scale to human clinical or population studies. Examples of co-analysis of "-omics" data include integrating metabolic profiles with quantitative trait locus data in a diabetic rat model (Dumas et al. 2007), combining metabolic and proteomic data for a mouse model of prostate cancer (Rantalainen et al. 2006). Bayesian methods for establishing correlations between two disparate sets of data show promise and have the added advantage of being nonlinear. Preliminary studies linking clinically measured lipoprotein measurements to ^{1}H NMR spectra enable some resolution of the highly overlapped lipoproteins in the NMR plasma spectra (Vehtari et al. 2007).

Metabolic profiling technology has come a long way since its origins in small-scale animal studies looking at gross metabolic changes. It is now an exquisitely sensitive tool for profiling multiple dynamic biological processes and cannot only accommodate the high degree of

Human Metabotype Profiling

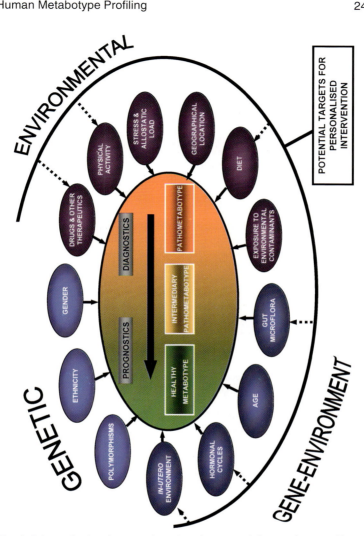

Fig. 4. Schematic showing genetic and environmental factors that contribute to the global metabolic profile of humans

metabolic variation typical of human data sets, but can help to unravel the various contributions from a range of genetic, epigenetic and environmental influences (Fig. 4). We are now standing on the threshold of a new era as this technology comes of age and is increasingly applied in the systems biology arena. With judicious application, this technology promises to deliver advances both in personalized health care and in population screening.

References

Atherton HJ, Bailey NJ, Zhang W, Taylor J, Major H, Shockcor J, Clarke K, Griffin JL (2006) A combined ^{1}H NMR spectroscopy- and mass spectrometry-based metabolomic study of the PPAR-alpha null mutant mouse defines profound systemic changes in metabolism linked to the metabolic syndrome. J Physiol Genomics 27:178–186

Atkinson AJ, Colburn WA, DeGruttola VG, DeMets DL, Downing GJ, Hoth DF, Oates JA, Peck CC, Schooley RT, Spilker BA et al. Biomarkers and surrogate end points: preferred definitions and conceptual framework

Bailey NJ, Sampson J, Hylands PJ, Nicholson JK, Holmes E (2002) Multicomponent metabolic classification of commercial feverfew preparations via high-field ^{1}H-NMR spectroscopy and chemometrics. Planta Med 68:734–738

Bales JR, Higham DP, Howe I, Nicholson JK, Sadler PJ (1984) Use of high-resolution proton nuclear magnetic resonance spectroscopy for rapid multicomponent analysis of urine. Clin Chem 30:426–432

Bates GP, Mangiarini L, Mahal A, Davies SW (1997) Transgenic models of Huntington's disease. Hum Mol Genet 6:1633–1637

Bollard ME, Holmes E, Lindon JC, Mitchell SC, Branstetter D, Zhang W, Nicholson JK (2001) Investigations into biochemical changes due to diurnal variation and estrus cycle in female rats using high resolution ^{1}H NMR spectroscopy of urine and pattern recognition. Anal Biochem 295:194–202

Brindle JT, Antti H, Holmes E, Tranter G, Nicholson JK, Bethell HW, Clarke LS, Schofield PM, McKilligin E, Mosedale DE, Grainger DJ (2002) Rapid and noninvasive diagnosis of the presence and severity of coronary heart disease using ^{1}H-NMR-based metabonomics. Nat Med 8:1439–1444

Buck AA, Anderson RI, MacRae AA (1978) Epidemiology of poly-parasitism. I. Occurrence, frequency and distribution of multiple infections in rural communities in Chad, Peru, Afghanistan, and Zaire. Tropenmed Parasitol 29:61–70

Clayton TA, Lindon JC, Cloarec O, Antti H, Charuel C, Hanton G, Provost JP, Le-Net JL, Baker D, Walley RJ, Everett JR, Nicholson JK (2006) Pharmaco-metabonomic phenotyping and personalized drug treatment. Nature 440: 1073–1075

Cloarec O, Dumas ME, Craig A, Barton RH, Trygg J, Hudson J, Blancher C, Gauguier D, Lindon JC, Holmes E, Nicholson J (2005) Statistical total correlation spectroscopy: an exploratory approach for latent biomarker identification from metabolic 1H-NMR data sets. Anal Chem 77:1282–1289

Coen M, O'Sullivan M, Bubb WA, Kuchel PW, Sorrell T (2005) Proton nuclear magnetic resonance-based metabonomics for rapid diagnosis of meningitis and ventriculitis. Clin Infect Dis 41:1582–1590

Crockford DJ, Holmes E, Lindon JC, Plumb RS, Zirah S, Bruce S, Rainville P, Stumpf CL, Nicholson JK (2006) Statistical heterospectroscopY (SHY), a new approach to the integrated analysis of NMR, UPLC-MS data sets: application in metabonomic toxicology studies. Anal Chem 78:363–371

Csenki L, Alm E, Torgrip RJ, Aberg RJ, Nord LI, Schuppe-Kostinen I, Lindberg J (2007) Proof of principle of a generalized fuzzy Hough transform approach to peak alignment of one dimensional ^{1}H NMR data. Anal Bioanal Chem 389:8775–885

Denkert C, Budczies J, Kind T, Weichert W, Tablack P, Sehouli J, Niesporek S, Könsgen D, Dietel M, Fiehn O (2006) Mass spectrometry-based metabolic profiling reveals different metabolite patterns in invasive ovarian carcinomas and ovarian borderline tumors. J Mass Spectrom 41:1546–1553

Dieterle F, Ross A, Schlotterbeck G, Senn H (2006) Probabalistic quotient normalization as a robust method to account for dilution of complex biological mixtures. Application in ^{1}H NMR metabonomics. Anal Chem 78:4281–4290

Dumas M-E, Barton RH, Toye A, Cloarec O, Craig A, Blancher C, Rothwell A, Fearnside J, Tatoud R, Blanc V, Lindon JC, Mitchell S, Holmes E, McCarthy MI, Scott J, Gaugier D, Nicholson JK (2006a) Metabolic profiling reveals a contribution of gut microbiota to insulin resistance phenotype in mice. Proc Natl Acad Sci U S A 103:12511–12516

Dumas ME, Maibaum EC, Teague C, Ueshima H, Zhou BF, Lindon JC, Nicholson JK, Stamler J, Elliott P, Chan Q, Holmes E (2006b) Assessment of analytical reproducibility of ^{1}H NMR spectroscopy based metabonomics for large-scale epidemiological research: the INTERMAP study. Anal Chem 78:2199–2208

Dumas ME, Wilder SP, Bihoreau MT, Barton RH, Fearnside JF, Argoud K, D'Amato L, Wallis RH, Blancher C, Keun HC, Baunsgaard D, Scott J, Sidelmann UG, Nicholson JK, Gauguier D (2007) Direct quantitative trait locus mapping of mammalian metabolic phenotypes in diabetic and normo-glycemic rat models. Nat Genet 39:666–672

Ebbels TM, Keun HC, Beckonert OP, Bollard ME, Lindon JC, Holmes E, Nicholson JK (2007) Prediction and classification of drug toxicity using probabilistic modeling of temporal metabolic data: the consortium on meta-bonomic toxicology screening approach. J Proteome Res 6:4407–4422

Fiehn O, Kopka J, Dormann P, Altmann T, Trethewey RN, Willmjtzer L (2000) Metabolite profiling for plant functional genomics. Nat Biotechnol 18: 1157–1161

Fossel ET, Carr JM, McDonagh J (1986) Detection of malignant tumors. Water-suppressed proton nuclear magnetic resonance spectroscopy of plasma. N Engl J Med 315:1369–1376

Foxall PJ, Lenz EM, Lindon JC, Neild GH, Wilson ID, Nicholson JK (1996) Nuclear magnetic resonance and high-performance liquid chromatography-nuclear magnetic resonance studies on the toxicity and metabolism of ifos-famide. Ther Drug Monit 18:498–505

Garrod SL, Humpfer E, Spraul M, Connor SC, Polley S, Connelly J, Lindon JC, Nicholson JK, Holmes E (1999) High resolution magic-angle-spinning ^{1}H NMR spectroscopic studies on intact rat renal cortex and medulla. Magn Res Med 41:1108–1118

Gavaghan CL, Holmes E, Lenz E, Wilson ID, Nicholson JK (2000) An NMR-based metabonomic approach to investigate the biochemical consequences of genetic strain differences: application to the C57BL10J, Alpk:ApfCD mouse. FEBS Lett 484:169–174

Glickman SE, Kilburn JO, Butler WR, Ramos LS (1994) Rapid identification of mycolic acid patterns of mycobacteria by high-performance liquid chro-matography using pattern recognition software and a *Mycobacterium* li-brary. J Clin Microbiol 32:740–745

Gruenewald TL, Seeman TE, Ryff CD, Singer BH (2006) Early warning bio-markers: what combinations predict later life mortality? Proc Natl Acad Sci 103:14158–14163

Guthrie R, Susi A (1963) A simple phenylalanine method for detecting phenyl-ketonuria in large populations of newborn infants. Pediatrica 132:328–343

Holmes E, Tsang TM, Huang JTJ, Leweke M, Koethe D, Gerth CW, Nolden BM Gross S, Schreiber D, Nicholson JK, Bahn S (2006) Metabolic profiling of CSF: a new tool for monitoring schizophrenia and response to therapeutic intervention. J PLOS Med 3:1420–1428

Holmes E, Foxall PJD, Nicholson JK, Neild GH, Brown SM, Beddell CR, Sweatman BC, Rahr E, Lindon JC, Spraul M, Neidig P (1994) Automatic data reduction and pattern recognition methods for analysis of ^{1}H nuclear magnetic resonance spectra of human urine from normal and pathological states. Anal Biochem 220:284–296

Holmes E, Loo RL, Cloarec O, Coen M, Tang H, Maibaum E, Bruce S, Chan Q, Elliott P, Stamler J, Wilson ID, Lindon JC, Nicholson JK (2007) Detectionof urinary drug metabolite (xenometabolome) signatures in molecular epidemiology studies via statistical total correlation spectroscopy. Anal Chem 79:2629–2640

Holmes E, Loo RL, Stamler J, Bictash M, Yap IK, Chan Q, Ebbels T, De Iorio M, Brown IJ, Veselkov KA, Daviglus ML, Ueshima H, Nicholson JK, Elliott P (2008) Human metabolic phenotype diversity and its association with diet and blood pressure. Nature 453:396–400

International Huntington Association and the World Federation of Neurology Research Group on Huntington's chorea (1994) Guidelines for the molecular genetics predictive test in Huntington's disease. J Med Genet. 31:555–559

Jenkins BG, Koroshetz WJ, Beal MF, Rosen BR (1993) Evidence for impairment of energy metabolism in vivo in Huntington's disease using localized ^{1}H NMR spectroscopy. Neurology 43:2689–2695

Jonsson P, Bruce SJ, Moritz T, Trygg J, Sjostrom M, Plumb R, Granger J, Maibaum E, Nicholson JK, Holmes E, Antti H (2005) Extraction, interpretation and validation of information for comparing samples in metabolic LC/MS data sets. Analyst 130:701–707

Jonsson P, Johansson ES, Wuolikainen A, Lindberg J, Schuppe-Koistinen I, Kusano M, Sjöström M, Trygg J, Moritz T, Antti H (2006) Predictive metabolite profiling applying hierarchical multivariate curve resolution to GC-MS data – a potential tool for multi-parametric diagnosis. J Proteome Res 5:1407–1414

Kirschenlohr HL, Griffin JL, Clarke SC, Rhydwen R, Grace AA, Schofield PM, Brindle KM, Metcalfe JC Proton (2006) NMR analysis of plasma is a weak predictor of coronary artery disease. Nat Med 12:705–710

Koschorek F, Offermann W, Stelten J, Braunsdorf WE, Stellar U, Gremmel H, Liebfritz D (1993) High resolution ^{1}H NMR spectroscopy of cerebrospinal fluid in spinal disease. Neurosurg Rev 16:302–315

Lamers RJ, DeGroot J, Spies-Faber EJ, Jellema RH, Kraus VB, Verzijl N, TeKoppele JM, Spijksma GK, Vogels JT, van der Greef J, van Nesselrooij JH (2003) Identification of disease- and nutrient-related metabolic fingerprints in osteoarthritic Guinea pigs. J Nutr 133:1776–1780

Lenz EM, Bright J, Wilson ID, Morgan SR, Nash AF (2003) A ^{1}H NMR-based metabonomic study of urine and plasma samples obtained from healthy human subjects. J Pharm Biomed Anal 33:1103–1115

Maher AD, Zirah SF, Holmes E, Nicholson JK (2007) Experimental and analytical variation in human urine in ^{1}H NMR spectroscopy based metabolic phenotyping studies. Anal Chem 79:5204–5211

Marchesi J, Holmes E, Khan F, Kochhar S, Scanalan P, Shanahan F, Wilson ID, Wang Y (2007) Rapid and noninvasive metabonomic characterization of inflammatory bowel disease. J Proteome Res 6:546–551

McEwen B (2002) Sex, stress and the hippocampus: allostasis, allostatic load and the aging process. Neurobiol Aging 23:921–939

Neild GH, Foxall PJ, Lindon JC, Holmes EH, Nicholson JK (1997) Uroscopy in the 21st century: high field NMR spectroscopy. Nephrol Dial Transplant 12:404–417

Nemoto T, Ando I, Kataoka T, Arifuku K, Kanazawa K, Natori Y, Fujiwara M (2007) NMR metabolic profiling combined with two-step principal component analysis for toxin-induced diabetes model rat using urine. J Toxicol Sci 32:429–435

Nicholson JK, Lindon JC, Holmes E (1999) 'Metabonomics': understanding the metabolic responses of living systems to pathophysiological stimuli via multivariate statistical analysis of biological NMR spectroscopic data. Xenobiotica 11:181–1189

Nicholson JK, Connelly J, Lindon JC, Holmes E (2002) Metabonomics: a generic platform for the study of drug toxicity and gene function. Nat Rev Drug Discov 1:153–161

Odunsi K, Wollman RM, Ambrosone CB, Hutson A, McCann SE, Tammela J, Geisler JP, Miller G, Sellers T, Cliby W, Qian F, Keitz B, Intengan M, Lele S, Alderfer JL (2005) Detection of epithelial ovarian cancer using H-1-NMR-based metabonomics. Int J Cancer 113:782–788

Perroud B, Lee J, Valkova N, Dhirapong A, Lin PY, Fiehn O, Kültz D, Weiss RH (2007) Pathway analysis of kidney cancer using proteomics and metabolic profiling. Anal Chem 79:6995–7004

Plumb RS, Stumpf CL, Granger JH, Castro-Perez J, Haselden JN, Dear GJ (2003) Use of liquid chromatography/time-of-flight mass spectrometry and multivariate statistical analysis shows promise for the detection of drug metabolites in biological fluids. Rapid Commun Mass Spectrom 17:2632–2638

Plumb RS, Granger JH, Stumpf CL, Johnson KA, Castro-Perez J, Wilson ID, Nicholson JK (2005) A rapid screening approach to metabonomics using UPLCand oa-TOF mass spectrometry: application to age, gender and diurnal variation in normal/Zucker obese rats and black and white nude mice. Analyst 130:844–849

Rantalainen M, Cloarec O, Beckonert O, Wilson ID, Jackson D, Tonge R, Rowlinson R, Rayner S, Nickson J, Wilknson RW, Mills JD, Trygg J, Nicholson JK, Holmes E (2006) Statistically integrated metabonomic-proteomic studies on human prostate cancer xenograft model in mice. J Proteome Res 5:2642–2655

Sabatine MS, Liu E, Morrow DA, Heller E, McCarroll R, Wiegand R, Berriz GF, Roth FP, Gerszten RE (2005) Metabolomic identification of novel biomarkers of myocardial ischemia. Circulation 112(25):3868–3875

Saric J, Wang Y, Li JV, Coen M, Utzinger J, Marchesi JR, Keiser J, Veselkov K, Lindon JC, Nicholson JK, Holmes E (2008) Species variation in the fecal metabolome gives insight into differential gastrointestinal function. J Proteome Res 7:352–360

Scanalan PD, Shanahan F, Clune Y, Collins JK, O'Sullivan GC, O'Riordan M, Holmes E, Wang Y, Marchesi J (2008) Culture-independent analysis of the gut microbiota in colorectal cancer and polyposis. Env Microbiol 10:789–798

Singer BH, Utzinger J, RYff CD, Wang Y, Holmes E (2006) Exploiting the potential of metabonomics in large population studies: three venues. In: Lindon JC, Nicholson JK, Holmes E (eds) The handbook of metabonomics and metabolomics, 1st edn. Elsevier, pp. 289–327

Slupsky CM, Rankin KN, Wagner J, Fu H, Chang D, Weljie AM, Saude EJ, Lix B, Adamko DJ, Shah S, Greiner R, Sykes BD, Marrie TJ (2007) Investigations of the effects of gender, diurnal variation and age in human urinary metabolomic profiles. Anal Chem 79:6995–7004

Stamler J, Elliott P, Dennis B, Dyer A, Kesteloot H, Liu K, Ueshima H, Zhou BF, for the INTERMAP Research Group. (2003) INTERMAP: background, aim, design, methods and descriptive statistics (non-dietary). J Hum Hypertension 17:591–608

Stoyanova R, Nicholson JK, Lindon JC, Brown TR (2004) Sample classification based on Bayesian spectral decomposition of metabonomic NMR data sets. Anal Chem 76:3666–3674

Teague C, Holmes E, Maibaum E, Nicholson J, Tang H, Chan Q, Elliot P, Wilson I (2004) Ethyl glucoside in human urine following dietary exposure: detection by ^{1}H NMR spectroscopy as a result of metabonomic screening of humans. Analyst 129:259–264

Teague CR, Dhabhar FS, Beckwith-Hall B, Holmes E, Powell J, Cobain M (2006) Metabonomic studies on the physiological effects of acute and chronic psychological stress in Sprague-Dawley rats. J Proteome Res 6:2080–2093

Teahan O, Gamble S, Holmes E, Waxman J, Nicholson JK, Bevan C, Keun HC (2006) Impact on analytical bias in metabonomic studies of human blood serum and plasma. Anal Chem 78:4307–4318

Tsang TM, Griffin JL, Haselden J, Holmes E (2005) Metabolic characterization of distinct neuroanatomical regions in rats by magic angle spinning H-1 nuclear magnetic resonance spectroscopy. Magn Res Med 53:1018–1024

Tsang TM, Huang JT, Holmes E, Bahn S (2006a) Metabolic profiling of plasma from discordant schizophrenia twins: correlation between lipid signals and global functioning in female schizophrenia patients. J Proteome Res 5:756–760

Tsang TM, Woodman B, McLoughlin GA, Griffin JL, Tabrizi SJ, Bates GP Holmes E (2006b) Metabolic characterization of the R6/2 transgenic mouse model of Huntington's disease by high-resolution MAS ^{1}H NMR spectroscopy. J Proteome Res 5:483–492

Underwood BR, Broadhurst D, Dunn WB, Ellis DI, Michell AW, Vacher C et al. (2006) Huntington disease patients and transgenic mice have similar procatabolic serum metabolite profiles. Brain 129:877–886

Vehtari A, Mäkinen VP, Soininen P, Ingman P, Mäkelä SM, Savolainen MJ, Hannuksela ML, Kaski K, Ala-Korpela M (2007) A novel Bayesian approach to quantify clinical variables and to determine their spectroscopic counterparts in ^{1}H NMR metabonomic data. BMC Bioinformatics 8 [Suppl 2]:S8

Vilasi A, Cutillas PR, Maher AD, Zirah SF, Capasso G, Norden AW, Holmes E, Nicholson JK, Unwin RJ (2007) Combined proteomic and metabonomic studies in three genetic forms of the renal Fanconi syndrome. Am J Physiol Renal Physiol 293:456–467

Wang Y, Holmes E, Comelli E, Fotopoulos G, Dorta G, Tang H, Rantalainen M, Lindon JC, Corthesy-Theulaz Fay LB, Kochhar S, Nicholson JK (2008) Topographical variation in metabolic signatures of human gut epithelial biopsies revealed by high-resolution magic-angle-spinning ^{1}H NMR spectroscopy. J Proteome Res 6:3944–3951

Wang YL, Tang HR, Nicholson JK et al. (2004) Metabolomic strategy for the classification and quality control of phytomedicine: a case study of chamomile flower (*Matricaria recutita* L) Planta Medica 70:250–255

Williams R, Lenz EM, Evans JA, Granger JH, Plumb RS, Stumpf CL (2005) A combined ^{1}H NMR, HPLC-MS-based metabonomics study of urine from obese (fa/fa) Zucker and normal Wistar-derived rats. J Pharm Biomed Anal 38:465–471

Williams RE, Lenz EM, Rantalainen M, Wilson ID (2006) The comparative metabonomics of age-related changes in the urinary composition of male Wistar-derived and Zucker (fa/fa) obese rats. Mol BioSystems 2:193–202

Yang C, Richardson AD, Smith JW, Osterman A (2007) Comparative metabolomics of breast cancer. Pac Symp Biocomput 181–192

Yi LZ, He J, Liang YZ, Yuan DL, Chau FT (2006) Plasma fatty acid metabolic profiling and biomarkers of type 2 diabetes mellitus based on GC/MS, PLS-LDA. FEBS Lett 580:6837–6845

Ernst Schering Foundation Symposium Proceedings, Vol. 4, pp. 251–264
DOI 10.1007/2789_2008_097
© Springer-Verlag Berlin Heidelberg
Published Online: 03 July 2008

Defining Personal Nutrition and Metabolic Health Through Metabonomics

S. Rezzi, F-P.J. Martin, S. Kochhar[✉]

BioAnalytical Science, Metabonomics and Biomarkers, Nestlé Research Center,
PO Box 44, 1000 Lausanne 26, Switzerland
email: *sunil.kochhar@rdls.nestle.com*

1	Metabonomics Technology Set-up	252
2	Nutrimetabonomics	252
3	The Complex Host–Microbiome Interactions: A Challenge for Nutrition	256
4	Characterizing the Metabolic Status of Individuals: A Step Toward Personalized Nutrition	260
References		261

Abstract. A major charter for modern nutrition is to provide a molecular basis for health outcome resulting from different food choices and how this could be designed to maintain individual health free of disease. Nutrigenomic techniques have been developed to generate information at various levels of biological organization, i.e. genes, proteins, and metabolites. Within this frame, metabonomics targets the molecular characterization of a living system through metabolic profiling. The metabolic profiles are explored with sophisticated data mining techniques mainly based on multivariate statistics, which can recover key metabolic information to be further linked to biochemical processes and physiological events. The power of metabonomics relies on its unique ability to assess functional changes in the metabolism of complex organisms stemming from multiple influences such as lifestyle and environmental factors. In particular, metabolic profiles encapsulate information on the metabolic activity of symbiotic partners, i.e. gut microflora, in complex organisms, which represent a major determinant in nutrition and health. Therefore, applications of metabonomics to nutrition sciences led to the nutrimetabonomics approach for the clas-

252   S. Rezzi et al.

sification of dietary responses in populations and the possibility of optimized or personalized nutritional management.

1   Metabonomics Technology Set-up

High-resolution proton nuclear magnetic resonance (^{1}H NMR) spectroscopy is an efficient and non destructive tool for generating data on a multitude of metabolites in biofluids or tissues. However, it is inherently less sensitive than mass spectrometry. When coupled to a liquid chromatography system, mass spectrometry (MS) provides a rapid platform for metabolite profiling at a concentration range of a few nM to pM. With the advent of ultraperformance liquid chromatography (Acquity UPLC system, Waters, Milford, MA, USA) combined with a time-of-flight mass spectrometer (MicroMass LCT-primers, Waters, Milford, MA, USA) equipped with an electrospray interface, complementary data to ^{1}H NMR profiling can be generated in 15–30 min per sample, thus enlarging the metabolite window for biomarker extraction (see Fig. 1 for a typical metabonomic analytical platform). The acquired spectral profile of a biofluid such as urine, plasma or saliva reflects the metabolic status of a living organism. Information recovery, in terms of relationships between the NMR, MS spectral profiles and their biochemical interpretation, can be maximized by applying multivariate statistical tools to analyze the information-rich spectroscopic data (see Fig. 2 for a typical data workflow). ^{1}H NMR and/or MS spectroscopy of complex biological mixtures combined with multivariate statistical analysis provides better visualization of the changing endogenous metabolite profile in response to physiological challenge or stimulus such as a disease process, administration of a xenobiotic, environmental stress, genetic modification, changes in nutrition and other physiological effects.

2   Nutrimetabonomics

Over the past decade, metabonomics has been shown to be a powerful tool to assess metabolic effects associated with toxicological insults (Coen et al. 2007; Ebbels et al. 2007; Skordi et al. 2007; Yap et al.

Fig. 1. Schematic drawing describing typical NMR- and MS-based metabonomic analytical platform

2006) and pathophysiological states (Brindle et al. 2002, 2003; Odunsi et al. 2005; Wang et al. 2005; Yang et al. 2004). Nutrimetabonomics provides a systems approach to characterize metabolic health and phenotypes of individuals. An individual human phenotype is determined by a complex interplay between genes, environmental and lifestyle factors, as well as intestinal symbionts (Gavaghan et al. 2004; Martin et al. 2007a; Nicholson et al. 2004; Nicholson and Wilson 2003). Recently, nutrimetabonomic strategies were successfully applied to classify dietary responses in populations and personalized nutritional management (Rezzi et al. 2007a).

Application of metabonomics to nutrition sciences, i.e. nutrimetabonomics, attempt to decipher metabolic effects of specific ingredients or diets in healthy human populations. This makes the task of nutrimetabonomics even more challenging than pharmaceutical or clinical metabo-

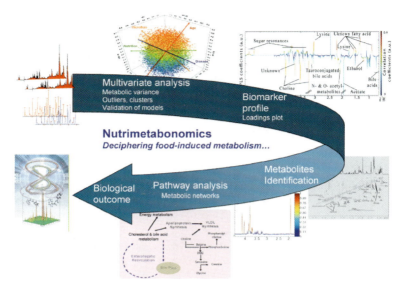

Fig. 2. Schematic description of a typical metabonomics data analysis workflow

nomics, since food-induced metabolic modulations appear as a result of complex interactions between the food and consumer metabolomes. In nutrition, effects indeed cannot be reduced to the effect of a single active molecule and are of low amplitude when compared to toxicological and clinical stressors. An additional source of complexity is attributable to inherent intra- and inter-individual variability that is reflected in the metabolism as a consequence of circadian rhythms, ovarian cycle, genetic polymorphism, different gut microflora activity, environmental and lifestyle components, and age. These variability sources, seen as confounding factors, can led to artifactual interpretation and therefore need to be controlled through an appropriate experimental design. In terms of analytical strategy, nutrimetabonomics requires well-established technical approaches most of the time based on NMR spectroscopy and MS in combination with multivariate statistics (Lindon et al. 2006).

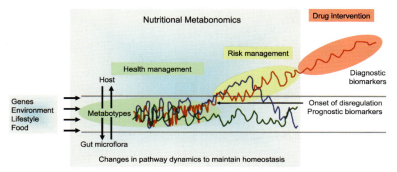

Fig. 3. Schematic representation of nutritional metabonomics concept. Different individual metabolic phenotypes (represented by *colored lines*) are under homeostatic controls maintaining metabolic variability within a healthy range (*green line*). Metabolic profiling will enable the prediction of individual susceptibility to develop well-defined diseases and optimizing nutrition for health maintenance and to restore homeostasis, as illustrated by the *blue line*. (Reproduced with permission from Rezzi et al. 2007a)

The sheer complexity of the web of interactions between host, gut microorganisms and the complex food metabolome is an important determinant to the response of the organism to a stressor or intervention that can easily be captured using metabolic profiling (Fig. 3). We have exemplified this nutrimetabonomic concept in a recent study where spectroscopically generated metabolic phenotypes obtained from healthy human volunteers were correlated with behavioral or psychological dietary preferences (Rezzi et al. 2007b). Dietary preferences and habits, which are predominantly cultural in origin, affect the health of both individuals and populations. In this study, we reported that metabolic profiling of urine and plasma samples revealed differential profiles for lipoproteins, basal energy metabolism and human-gut–microbial metabolic interactions. The observed metabolic imprinting findings provide evidence for a link between specific dietary preferences and metabolic phenotypes in both human basal metabolism and gut microbial activity, which in turn may have long-term health consequences.

These data suggested that gut microbial metabolism in humans may be modulated by the diet more than previously thought, which was previously suggested in previous studies. For instance, the changes in the metabolism of human subjects caused by variations in their vegetarian, low-meat, and high-meat diets were investigated in a crossover design (Stella et al. 2006). Individuals consuming the vegetarian diet showed metabolic signatures indicative of an altered metabolic activity of gut bacteria with variations in the urinary levels of 4-hydroxyphenylacetate and p-cresol sulfate. The authors reported that conversion of 4-hydrohyphenylacetate to p-cresol was carried out by a limited number of microbial species and could be used to assess changes in gut bacterial activities in response to nutritional interventions. More recently, potential metabolic imprinting in the bacterial activities in response to aging and specific dietary intervention were exhaustively described in a long-term study on caloric restriction (CR) in dog (Wang et al. 2007). Metabolic profiling of urine samples revealed signatures related to aging and growth-related biological processes as well as CR-induced effects on basal energy metabolism. The authors highlighted that both aging and CR led to differential profiles of mammalian-microbial co-metabolites, such as aromatic derivates, i.e. hippurate, 3-hydroxy-phenylpropioniate, as well as aliphatic methylamines, which may indicate altered activities of the gut microbiome.

3   The Complex Host–Microbiome Interactions: A Challenge for Nutrition

The recent discovery of the contribution of gut microbiota in the predisposition to gastritis and obesity has raised new interest in gut microbial activities of human and their implications in future nutritional healthcare. The gut microbiome-mammalian superorganism (Lederberg 2000) represents a level of biological evolutionary development in which there is extensive transgenomic modulation of metabolism and physiology that is a characteristic of true symbiosis (Martin et al. 2007c). The gut microbial community contains multiple cell types providing an extended genome interacting with a number of important mammalian metabolic regulatory functions. As the microbiome interacts strongly

Nutrimetabonomics

with the host to determine the metabolic phenotype (Dumas et al. 2006; Martin et al. 2007a) and metabolic phenotype influences on the outcomes of drug interventions (Clayton et al. 2006; Nicholson et al. 2004), there is clearly an important role of understanding these interactions as part of personalized healthcare solutions. Several mammalian–microbial associations, both positive and negative, have already been studied with metabonomics (Dunne 2001; Gill et al. 2006; Ley et al. 2006; Nicholson et al. 2005; Verdu et al. 2004).

Recently, the contribution of the gut microflora on mammalian metabolism was revisited using metabolic profiling. The importance of the metabolic relationship between gut microflora and host physiology was demonstrated in a study assessing the effect of dietary changes, i.e. switching from a 5% control low-fat diet to a 40% high-fat diet, on the metabolic status of insulin-resistant mice (Dumas et al. 2006). In this experiment, mice showed a significant association between a specific metabolic phenotype, e.g. low plasma phosphatidylcholine and high urinary methylamines, and genetic predisposition to high-fat diet-induced dyslipidemia and nonalcoholic fatty liver. The urinary excretion of methylamines was directly related to microflora metabolism, suggesting that conversion of choline into methylamines by microbiota in this experimental design mimics the effect of choline-deficient diets, causing nonalcoholic fatty liver disease. These data also indicate that gut microbiota may play an active role in the development of insulin resistance.

The intimate relationship that animals have with the organisms inhabiting their guts is also very well illustrated by Martin et al., who describe the results of a top-down view of these microbial–mammalian interactions, showing the effects of different gut flora on the metabolic profiles of the mouse organism (Martin et al. 2007a). This study aimed at assessing the metabolic effects of inoculating germ-free mice with human-derived flora (nonadapted microflora) or exposing them to conventional mice (re-conventionalization), enabling the acquisition of a normal mouse microbiome. Metabolic profiling revealed that reconventionalized mice tend to converge metabolically and ecologically toward conventional mice with a healthier physiology. Inoculation of germ-free mice with a nonadapted microflora modifies the physiology of the murine host toward a prepathologic state and maintains the gut tract and the liver outside a sustainable mouse ecological equilibrium. It was shown that a nonadapted gut microbiota is critically involved in supplying host calorific requirement via reprocessing of bile acids. This is part of what Martin et al. have termed the microbiome–host metabolic axis, i.e. "the multi-way exchange and co-metabolism of compounds between the host organism and the gut microbiome resulting in transgenomically regulated secondary metabolites which have biological activity in both host and microbial compartments." Understanding the effects of bacterial metabolism on the balance of bile acids in enterohepatic recirculation is a major challenge due to the implications of microbiota in fat absorption, lipid metabolism, drug therapeutic or toxic effects as well as direct effects within the gastrointestinal tract and its contents. In that regard, recent advances in microbial and metabolic profiling make possible the multicompartment study of bile acids and their effects on intermediary metabolism.

Much attention has recently been focused on the use of probiotic supplements as a means to promote gut health, thus preventing the incidence of allergies and inflammatory states. A probiotic is generally defined as a "live microbial feed supplement added to appropriate food vehicles," which is expected to benefit intestinal microbial balance and consequently the host physiology (Gibson and Fuller 2000). In a follow-up study, Martin et al. have assessed the transgenomic metabolic effects of exposure to either *Lactobacillus paracasei* or *Lactobacillus rhamnosus* probiotics in this humanized microbiome mouse model (Martin et al. 2008). The authors have illustrated the robust capabilities of metabonomics to capture subtle metabolic fluctuations in diverse biological compartments including biofluids, tissues, and cecal content in relation to modulation of microbial population. The authors exemplified how multicompartmental transgenomic metabolic interactions could be resolved at the compartment and pathway level. They have described how probiotic exposure exerted microbiome modification with subsequent alteration of lipid metabolism and glycolysis. Probiotic treatments also altered a diverse range of pathways, including metabolism of amino acids, methylamines and short-chain fatty acids. These integrated system investigations demonstrate the usefulness of a top-down systems biology concept based on metabolic profiling to investigate the mechanistic bases of probiotics and the monitoring of their effects on the gut's microbial activity.

These investigations bring further evidence on the crucial role of the gut microbial activity on human health that have previously been raised in studies exploring the physiological effects of probiotic interventions in different mouse models. To begin with, the effects of a therapeutic intervention with *L. paracasei* probiotic on normalizing the metabolic disorders have been assessed in a model of *Trichinella spiralis*-induced irritable bowel syndrome (Martin et al. 2006). Both systemic and tissue-specific metabolic changes were captured using a metabonomic approach, which were consistent with an increase in energy requirement due to muscular hypercontractility and hypertrophy, inflammation and alteration of gut microbial activities. The authors illustrated partial regression of the metabolic disorders achieved by intervention with a *L. paracasei* probiotic. The probiotic treatment moved the energetic metabolism toward normality, reduced the gut microbiota disturbances and might contribute to normalization of the late inflammatory markers.

In other investigations, we have also explored the role of single probiotic inoculation on physiological status of different sections of the gastrointestinal tract of animals that were raised without any resident microorganisms (Martin et al. 2007b). Metabolic signatures reflecting the structure and function of the different compartments were obtained with variations in concentrations of amino acids, antioxidants, osmolytes and creatine. For instance, jejunum and colon showed metabolic signatures ascribed to lipogenesis and fat storage. More interestingly, metabonomic allowed capturing region-dependent metabolic changes triggered by ingestion of live *L. paracasei* in the upper gut, consistent with modulation of intestinal digestion, absorption, amino acid homeostasis, lipid metabolism and protection against oxidative stress. Contrary to the effects induced by live *L. paracasei*, no changes were seen with supplementation of irradiation-killed bacteria, which suggested that the differential metabolism observed with live bacteria is probably due to genuine host–bacteria interactions.

4 Characterizing the Metabolic Status of Individuals: A Step Toward Personalized Nutrition

As the microbiome interacts strongly with the host to determine the metabolic phenotype, metabolic health and nutritional status, there is clearly an important role of understanding these interactions as part of optimized nutrition. Nutrimetabonomics provides a system approach that is potentially able to assess metabolic status of individuals considering their specificity in terms of genetic and environmental factors, gut microbiota activity, lifestyle and food habits.

The characterization of the metabotype of individuals could open access to important information on dietary variations in humans and on the degree of response to dietary modulations. This may ultimately provide new insights into the role of diet and nutrition for health maintenance and personalized healthcare nutrition programs. Personalization of nutrition is the outcome for individuals who will adapt their diet and lifestyle according to knowledge about their current and future health status, and their subsequent nutritional requirements (see Fig. 4 for the conceptualization of nutritional metabonomics for health and risk management). This implies the development of analytical strategies leading to the characterization of the initial nutritional status on which diet or lifestyle recommendations could be applied to maintain or even improve metabolic health. In this way, the concept of "pharmaco-metabonomics" developed by Clayton et al. is interesting because providing a means to predict the metabolic response of a living organism from a simple preintervention metabolic profile (Clayton et al. 2006). Transposed to nutrition, such a concept could be used to optimize dietary recommendations for individuals. In order to achieve this goal, information contained in complex metabolic profiles would need to be validated for well-determined physiological and nutritional outcomes. In such cases, a dietetics professional could use this metabolic information to develop coordinated approaches to optimize and maintain metabolic health, proposing nutritional solutions consistent with the metabotype of individuals considering their lifestyle and health aspirations.

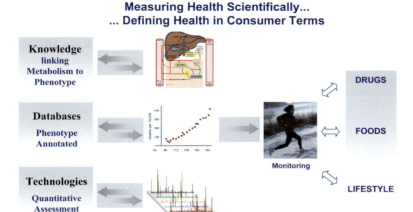

Fig. 4. Metabolic profiling as an approach to personalized health and nutrition

References

Brindle JT, Antti H, Holmes E, Tranter G, Nicholson JK, Bethell HW, Clarke S, Schofield PM, McKilligin E, Mosedale DE, Grainger DJ (2002) Rapid and noninvasive diagnosis of the presence and severity of coronary heart disease using 1H-NMR-based metabonomics. Nat Med 8:1439–1444

Brindle JT, Nicholson JK, Schofield PM, Grainger DJ, Holmes E (2003) Application of chemometrics to 1H NMR spectroscopic data to investigate a relationship between human serum metabolic profiles and hypertension. Analyst 128:32–36

Clayton TA, Lindon JC, Cloarec O, Antti H, Charuel C, Hanton G, Provost JP, Le Net JL, Baker D, Walley RJ, Everett JR, Nicholson JK (2006) Pharmacometabonomic phenotyping and personalized drug treatment. Nature 440:1073–1077

Coen M, Hong YS, Clayton TA, Rohde CM, Pearce JT, Reily MD, Robertson DG, Holmes E, Lindon JC, Nicholson JK (2007) The mechanism of galactosamine toxicity revisited; a metabonomic study. J Proteome Res 6:2711–2719

Dumas ME, Barton RH, Toye A, Cloarec O, Blancher C, Rothwell A, Fearnside J, Tatoud R, Blanc V, Lindon JC, Mitchell SC, Holmes E, McCarthy MI, Scott J, Gauguier D, Nicholson JK (2006) Metabolic profiling reveals a contribution of gut microbiota to fatty liver phenotype in insulin-resistant mice. Proc Natl Acad Sci U S A 103:12511–12516

Dunne C (2001) Adaptation of bacteria to the intestinal niche: probiotics and gut disorder. Inflamm Bowel Dis 7:136–145

Ebbels TM, Keun HC, Beckonert OP, Bollard ME, Lindon JC, Holmes E, Nicholson JK (2007) Prediction and classification of drug toxicity using probabilistic modeling of temporal metabolic data: the consortium on metabonomic toxicology screening approach. J Proteome Res 6:4407–4422

Gavaghan CL, Holmes E, Lenz E, Wilson ID, Nicholson JK (2004) An NMR-based metabonomic approach to investigate the biochemical consequences of genetic strain differences; application to the C57BL10J, Alpk:ApfCD mouse. FEBS Lett 484:169–174

Gibson GR, Fuller R (2000) Aspects of in vitro and in vivo research approaches directed toward identifying probiotics and prebiotics for human use. J Nutr 130:391S–395S

Gill SR, Pop M, Deboy RT, Eckburg PB, Turnbaugh PJ, Samuel BS, Gordon JI, Relman DA, Fraser-Liggett CM, Nelson KE (2006) Metagenomic analysis of the human distal gut microbiome. Science 312:1355–1359

Lederberg J (2000) Infectious history. Science 288:287–293

Ley R, Turnbaugh P, Klein S, Gordon J (2006) Microbial ecology: human gut microbes associated with obesity. Nature 444:1022–1023

Lindon JC, Holmes E, Nicholson JK (2006) Metabonomics techniques and applications to pharmaceutical research and development. Pharm Res 23:1075–1088

Martin FP, Verdu EF, Wang Y, Dumas ME, Yap IK, Cloarec O, Bergonzelli GE, Corthesy-Theulaz I, Kochhar S, Holmes E, Lindon JC, Collins SM, Nicholson JK (2006) Transgenomic metabolic interactions in a mouse disease model: interactions of *Trichinella spiralis* infection with dietary *Lactobacillus paracasei* supplementation. J Proteome Res 5:2185–2193

Martin FP, Dumas ME, Wang Y, Legido-Quigley C, Yap IK, Tang H, Zirah S, Murphy GM, Cloarec O, Lindon JC, Sprenger N, Fay LB, Kochhar S, van BP, Holmes E, Nicholson JK (2007a) A top-down systems biology view of microbiome-mammalian metabolic interactions in a mouse model. Mol Syst Biol 3:112

Martin FP, Wang Y, Sprenger N, Holmes E, Lindon JC, Kochhar S, Nicholson JK (2007b) Effects of probiotic *Lactobacillus paracasei* treatment on the host gut tissue metabolic profiles probed via magic-angle-spinning NMR spectroscopy. J Proteome Res 6:1471–1481

Nutrimetabonomics

Martin FP, Wang Y, Sprenger N, Yap IK, Lundstedt T, Lek P, Rezzi S, Ramadan Z, van Bladeren P, Fay LB, Kochhar S, Lindon JC, Holmes E, Nicholson JK (2008) Probiotic modulation of symbiotic gut microbial-host metabolic interactions in a humanized microbiome mouse model. Mol Syst Biol 4:157

Nicholson JK, Wilson ID (2003) Opinion: understanding 'global' systems biology: metabonomics and the continuum of metabolism. Nat Rev Drug Discov 2:668–676

Nicholson JK, Holmes E, Lindon JC, Wilson ID (2004) The challenges of modeling mammalian biocomplexity. Nat Biotechnol 22:1268–1274

Nicholson JK, Holmes E, Wilson ID (2005) Gut microorganisms, mammalian metabolism and personalized health care. Nat Rev Microbiol 3:431–438

Odunsi K, Wollman RM, Ambrosone CB, Hutson A, McCann SE, Tammela J, Geisler JP, Miller G, Sellers T, Cliby W, Qian F, Keitz B, Intengan M, Lele S, Alderfer JL (2005) Detection of epithelial ovarian cancer using 1H-NMR-based metabonomics. Int J Cancer 113:782–788

Rezzi S, Ramadan Z, Fay LB, Kochhar S (2007a) Nutritional metabonomics: applications and perspectives. J Proteome Res 6:513–525

Rezzi S, Ramadan Z, Martin FP, Fay LB, Bladeren PV, Lindon JC, Nicholson JK, Kochhar S (2007b) Human metabolic phenotypes link directly to specific dietary preferences in healthy individuals. J Proteome Res 6:4469–4477

Skordi E, Yap IK, Claus SP, Martin FP, Cloarec O, Lindberg J, Schuppe-Koistinen I, Holmes E, Nicholson JK (2007) Analysis of time-related metabolic fluctuations induced by ethionine in the rat. J Proteome Res 6:4572–4581

Stella C, Beckwith-Hall B, Cloarec O, Holmes E, Lindon JC, Powell J, vanderOuderaa F, Bingham S, Cross AJ, Nicholson JK (2006) Susceptibility of human metabolic phenotypes to dietary modulation. J Proteome Res 5:2780–2788

Verdu EF, Bercik P, Bergonzelli GE, Huang XX, Blennerhasset P, Rochat F, Fiaux M, Mansourian R, Corthesy-Theulaz I, Collins SM (2004) *Lactobacillus paracasei* normalizes muscle hypercontractility in a murine model of postinfective gut dysfunction. Gastroenterology 127:826–837

Wang C, Kong H, Guan Y, Yang J, Gu J, Yang S, Xu G (2005) Plasma phospholipid metabolic profiling and biomarkers of type 2 diabetes mellitus based on high-performance liquid chromatography/electrospray mass spectrometry and multivariate statistical analysis. Anal Chem 77:4108–4116

Wang Y, Lawler D, Larson B, Ramadan Z, Kochhar S, Holmes E, Nicholson JK (2007) Metabonomic investigations of aging and caloric restriction in a lifelong dog study. J Proteome Res 6:1846–1854

Yang J, Xu G, Hong Q, Liebich HM, Lutz K, Schmulling RM, Wahl HG (2004) Discrimination of type 2 diabetic patients from healthy controls by using metabonomics method based on their serum fatty acid profiles. J Chromatogr B Analyt Technol Biomed Life Sci 813:53–58

Yap IK, Clayton TA, Tang H, Everett JR, Hanton G, Provost JP, Le Net JL, Charuel C, Lindon JC, Nicholson JK (2006) An integrated metabonomic approach to describe temporal metabolic disregulation induced in the rat by the model hepatotoxin allyl formate. J Proteome Res 5:2675–2684

Ernst Schering Foundation Symposium Proceedings

Editors: Günter Stock
Monika Lessl

Vol. 1/2006: Tissue-Specific Estrogen Action
Editors: K.S. Korach, T. Wintermantel

Vol. 2/2006: GPCRs: From Deorphanization
to Lead Structure Identification
Editors: H.R. Bourne, R. Horuk, J. Kuhnke, H. Michel

Vol. 3/2006: New Avenues to Efficient Chemical Synthesis
Editors: P.H. Seeberger, T. Blume

Vol. 4/2006: Immunotherapy in 2020
Editors: A. Radbruch, H.-D. Volk, K. Asadullah, W.-D. Doecke

Vol. 5/2006: Cancer Stem Cells
Editors: O.D. Wiestler, B. Haendler, D. Mumberg

Vol. 1/2007: Progestins and the Mammary Gland
Editors: O. Conneely, C. Otto

Vol. 2/2007: Organocatalysis
Editors: M.T. Reetz, B. List, S. Jaroch, H. Weinmann

Vol. 3/2007: Sparking Signals
Editors: G. Baier, B. Schraven, A. von Bonin, U. Zügel

Vol. 4/2007: Oncogenes Meet Metabolism
Editors: G. Kroemer, D. Mumberg, H. Keun, B. Riefke,
T. Steger-Hartmann, K. Petersen

This series will be available on request from
Ernst Schering Foundation, Friedrichstraße 82, 10117 Berlin, Germany

Printing: Krips bv, Meppel, The Netherlands
Binding: Stürtz, Würzburg, Germany